Construction for a Regenerative Future

T0136398

This is a book about how to manage the processes involved in a construction project towards a sustainable and regenerative endproduct. It covers key project management concepts and links the construction process to the objectives of UN SDGs and beyond zero carbon emissions throughout the whole project life cycle.

This introductory textbook is written from a project manager's perspective including considerations of circular economy throughout the construction process focusing on a regenerative or restorative outcome. The book examines the importance of the type and purpose of a building, circularity and de-construction, the site, the client and its organisation, stakeholder considerations, the project organisation, the procurement of consultants and contractors, project performance during design and construction, project hand-over to the client, and the building's operation and maintenance. It also illustrates how to verify the building using existing environmental certifications, how to calculate carbon emissions, and how to deal with used construction materials from a circular economy perspective. International examples of best practice are included throughout, and the book is structured in a way which students will find engaging and easy to follow.

This is an ideal textbook for use on construction, architecture, and engineering programmes where the emphasis must urgently be placed on students fostering regenerative construction solutions in their coming professional life.

Urban Persson is a PhD and Lecturer in the Division of Construction Management at Lund University, Sweden. He has 35 years of experience from consultancy in construction and environmental management and as CEO of a real estate company.

Construction for a Regenerative Future

Urban Persson

Routledge
Taylor & Francis Group

LONDON AND NEW YORK

Cover credit: © shomos uddin/Getty Images

First published 2023
by Routledge
4 Park Square, Milton Park, Abingdon, Oxon OX14 4RN

and by Routledge
605 Third Avenue, New York, NY 10158

Routledge is an imprint of the Taylor & Francis Group, an informa business

© 2023 Urban Persson

British Library Cataloguing-in-Publication Data
A catalogue record for this book is available from the British Library

ISBN: 978-1-032-01210-0 (hbk)
ISBN: 978-1-032-00365-8 (pbk)
ISBN: 978-1-003-17770-8 (ebk)

DOI: 10.1201/9781003177708

Typeset in Times New Roman
by codeMantra

To my family

Contents

Figures

Foreword

Since the start of conscious construction management, the focus from the client's perspective has been on an efficient process. To control the process and reach the client's objectives, the project managers used different tools to manage the process towards the objective. The tools used were time schedules, bills of quantities, quality systems, etc. Although the construction process became very efficient (i.e. using fewer resources), the environmental impacts were often considerable.

Nowadays, most clients add some sustainability demands to the project objective. The project manager must use supplemental tools to manage the environmental impacts, particularly the construction process's contribution to carbon emissions. The tools used are based on Life Cycle Analysis (LCA), different certification schemes (LEED, BREEAM etc.), and a combination of the two mentioned. Construction projects with objectives including sustainability demands are not supposed to reach more than just sustainability. This construction activity level does not improve our planet's situation. In the best case, the environmental impact is zero.

A committed client could be interested in going beyond sustainability to a restorative or regenerative level when the construction process fulfils. One condition to reach such a successful project outcome is that the client, the project manager, architects and others in the project organization have the knowledge to go beyond sustainability. It is easy to understand that it is a strong need for a book dealing with the construction process for a regenerative future. I strongly recommend this textbook for the students, for all in the construction industry and our planet.

Bengt Hansson
Professor
Former head of the division of Construction management
Lund University

Preface

The construction sector is one of the main contributors to the negative environmental impact on our planet. Furthermore, it has a very long-term impact in comparison with most other industry sectors. To meet the zero-carbon emission goal set for 2050, it is crucial to considerably reduce the impact of the construction sector. The coming engineering and architect generations must accomplish most of this work. This subject is a very urgent concern for our common future. It is crucial to adapt these considerations before it is too late.

This book provides guidance on how to manage the process of a construction project with sustainability considerations towards a regenerative end product – a building – with inputs for the building's entire life cycle. The aim is to compile knowledge on sustainability and regenerative development from a construction process point of view.

Regenerative construction is the main objective if the construction sector expects to meet the requirements of zero carbon emissions and beyond zero before 2050 and thereafter. Considering this, the United Nations' Sustainable Development Goals (SDGs) are considered from a construction perspective during a building's entire life cycle. The book is written from a project manager's perspective and does not take special architectural considerations into account; however, it provides relevant examples of regenerative design approaches towards a regenerative construction process. The book examines the importance of the type and purpose of a building, the site, client and its organisation, stakeholder considerations, project organisation, design approaches, procurement of consultants and contractors, project handover to the client, the building's operation and maintenance process, how to verify the building using existing certifications of sustainability and regenerative building, and how to calculate carbon emissions.

Following a general introduction to the existing circumstances for the built environment, the starting point of the regenerative construction process begins. During the writing process of this textbook, it became obvious that a construction process that includes a circular economy is a prerequisite for sustainability and a regenerative outcome; thus, the book begins with the end of the construction process with considerations of deconstruction and disassembling. Consequently, it also illustrates how to deal with reused and reusable construction materials, whether harmful or not harmful for health and the environment, from a circular economy point of view. All these activities are from a sustainable and regenerative perspective.

The main purpose of this book is to serve as a reference for university engineering students worldwide, where English is spoken, to encourage regenerative solutions in construction in their professional lives. It is also suitable for practitioners interested in the future of regenerative issues.

The content of the book is primarily based on the latest scientific findings regarding sustainability and the built environment, regardless of country and region. However, it contains various Scandinavian input in places and is, of course, coloured with the author's interpretations. The content should not be considered the sole and complete roadmap to a regenerative construction process; the reader can seek other ways to plan and implement a regenerative building in the referenced literature. Many more sources of information can be found in the references.

Author background

I have been working with green and sustainable construction since the mid-1990s, both as a consultant in construction management and as a lecturer and teacher for university engineering students. I have followed 30 years of development in the environmental and sustainability movement of the construction sector, from single enthusiasts and ecovillage projects through the development of ecological and energy savings concerns from the end of the 1990s to the beginning of the 2000s, followed by the construction sector's recognition of the triple bottom-line of sustainability, climate change impacts, and carbon emission reduction. The next steps in this evolution seem to be zero carbon and energy renewable sources, a stringent circular economy, and roadmaps towards zero and beyond zero emissions and impacts on restorative and regenerative outcomes.

I completed my PhD in 2009 with the thesis 'Management of sustainability in construction works'. I have participated as a co-author in three volumes of Swedish textbooks for university engineering students in the subject of construction management (*Byggledning* in Swedish): the design stage (2015), the construction stage (2017), and the operation and maintenance stage (2020). My portion encompassed sustainability matters during the construction process. I have also attended, with one exception, all the international conference series of World Sustainable Built Environment since the beginning of 1998 in Vancouver, Canada, and I have contributed various papers during these conferences; the latest was held virtually (2020) in Gothenburg, Sweden, and the next will meet in 2023 in Montreal, Canada. In addition, I have worked as a site manager, contractor, and CEO for a Small and Medium Enterprises real estate company.

I would like to thank professor Anne Landin and professor Bengt Hansson at the divison of Construction Management, Lund University.

Finally, I want to give a lot of gratitude to my wife Gerd for her patience and perseverance and to my buddy Spencer, our unbelievable Irish Terrier dog.

Lund, Sweden
Urban Persson

1 Why – introduction

1.1 Sustainability and construction

The present global condition regarding climate and environment is a consequence of the level of human consumption of natural resources exceeding the earth's ability to sustain such use over the long term. The effects are degradation of eco-systems and severe conditions for human life. Buildings must be designed for resilience and adaptability to accommodate our changing global climate, especially in many developing countries, which are particularly susceptible to the effects of climate change. Not only buildings but the spaces in between (i.e. the infrastructure) must be equally sustainable and resilient to future risks. Approximately, $90 trillion USD of worldwide investment is needed until 2030 in the infrastructure sector to achieve a successful, net zero emissions future. Further, striving for buildings that push the boundaries on sustainability, such as those with net zero emissions, is a major driver for innovation and technology.

Sustainable development and the precautionary principle

Beginning in the 1960s (inspired by Rachel Carson's book *Silent Spring* in 1962), the term 'sustainable development' has been gradually defined by different interpretations. Since the beginning of the 1980s, the concept has entailed precautionary approaches, such as guidance measures for market failures, ensuring regenerative capacity of renewable resources, avoidance of pollution, and making product processes more eco-efficient. The 1987 UN report by the Brundtland commission conveys the most common and widely recognised definition of sustainable development under 'Our common future' (p. 24):

> *Humanity has the ability to make development sustainable to ensure it meets the needs of the present without compromising the ability of future generations to meet their own needs.*

Most of the varied interpretations of the term sustainable development share the following three main objectives to meet sustainability:

- Development should prioritise the remaining basic ecological processes: i.e. support life processes and prioritise human well-being.
- Development should protect biodiversity.
- Development should promote sustainable use of biodiversity and the entire environment over the long term.

DOI: 10.1201/9781003177708-1

Biodiversity refers to the variety of individuals that exist in a population of living species, the basis of natural selection, adaptation, and evolution.

During the 1992 UN Rio Summit (Rio de Janeiro, Brazil), Agenda 21 was established as a global roadmap to meet the requirements of sustainable development for the 21st century. The plan was categorised into three major areas – ecological, economic, and social development – known as the triple bottom-line.

Ecological development is significant to:

- Minimise emissions to air, ground, and water
- Conserve biodiversity
- Conserve finite resources
- Conserve resilience within ecosystems
- Protect ethical and cultural values associated with nature

Economic development is significant to promote

- Allocation of resources between countries and people
- Ability to work and to earn a living
- International trade that supports sustainability
- Sustainable production and consumption
- Technical development
- Corporate responsibility
- Consumer influence

Social development is significant to promote

- Quality of life and health
- Access to housing
- Education
- Possibility for individuals to influence the development of society and democracy
- Free access to information
- Ability to practice one's own culture
- Freedom of religion and speech
- Equality

Elements within the triple bottom-line should be considered on an equal level (see Figure 1.1) or as a 'Russian doll' model (see Figure 1.2). Figure 1.1 ensures that the solution is an optimisation problem with all three parts at an equal level. As the figure illustrates, the state of sustainability occurs in the middle where all three areas are interconnected. According to the 'Russian doll' model, the environment is considered the dominant part, and economic activity depends on social issues, both of which are constrained by environmental factors.

An alternate way of looking at sustainability is to consider the economic aspect of the triple bottom-line as the five capital model of sustainability. This implies that five capital assets are linked with the other parts of the triple bottom-line:

- *Financial capital* – Has no real value but represents natural, human, social, and manufactured capital, e.g. shares, bonds, and banknotes

Figure 1.1 The triple bottom-line of sustainable development (Persson, U. (2009)).

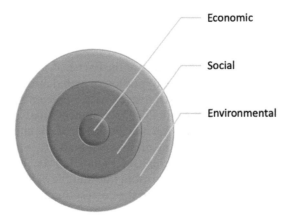

Figure 1.2 The Russian doll model of sustainability (Persson, U. (2009)).

- *Manufactured capital* – Commodities which contribute to the production process, e.g. tools, machines, and buildings
- *Social capital* – Institutions such as family, community, business, and unions that help maintain and develop human capital
- *Human capital* – Health, knowledge, skills, and motivation
- *Natural capital* – Resources (renewable and non-renewable), sinks (absorbs, neutralises, or recycled waste), and processes such as climate regulation.

Other dimensions, such as moral, technical, and political, are also mentioned. The implication is that these dimensions interact on different levels, with the moral dimension being the highest, followed by the original triple bottom-line elements on the next level and on the lowest level, the political, technical, and legal dimensions. However, this interaction is not due to activity between the dimensions; it is attributed more to the need to achieve solutions through holistic thinking, that is, between the complex problems of interconnected and interdependent relationships which determine the interactions within and between the dimensions. Considering the triple bottom-line's three original dimensions on an equal level (as in Figure 1.1) was the most common way to interpret sustainability in the construction sector until the UN sustainability goals were established (see below).

The 1992 UN Rio Summit also defined the precautionary principle (Principle 15) as:

> *To protect the environment, the precautionary approach shall be widely applied by states according to their capabilities. Where there are threats of serious or irreversible damage, lack of full scientific certainty shall not be used as a reason for postponing cost-effective measures to prevent environmental degradation.*

This principle is a highly significant aspect regarding the practical management of sustainability when scientific proof is interpreted differently or does not apply to the actual issue. For example, how radiation from electromagnetic fields originating from high-voltage wires in urban areas should be addressed. In some countries, the precautionary principle is regulated by environmental legislation.

UN sustainability goals and the construction sector

In 2015, the UN Climate Change Conference or Conference of the Parties, COP21, was held in Paris, where the new goal regarding the absolute limitation of global mean temperature was set to 2° Celsius (C) above the pre-industrial level, with a precautionary and desirable limit of 1.5°C. At COP21, the UN Sustainable Development Goals (UN SDGs) were established as a global goal until 2030, as Agenda 2030. The SDGs encompass 17 different goals relating to 17 different areas of urban sustainability, including no poverty, zero hunger, good health, and well-being (see Figure 1.3). Each goal aims to integrate the 'Russian Doll' model of the triple bottom line and the SDGs (see Figure 1.4), and the connection between specific SDGs and the corresponding part of the triple bottom-line.

The World Green Building Council (WGBC) is an independent and non-profit network organisation for businesses, governments, and organisations to promote and transform the building and construction sector to fulfil the UN SDGs. The organisation's network comprises approximately 70 different green building councils around the world that work in three strategic areas: climate action, health and well-being, and resources and circularity. With the approach of system change, the network is incentivising the construction industry to work towards creating net zero carbon, healthy, equitable, and resilient built environments by developing and using different available regional rating tools or certifications (see also Chapter 7).

Concerning the construction sector, the most important goal is number 11 – Creating sustainable cities and communities. Some examples of the targets are to ensure access to adequate, safe, and affordable housing; provide access to safe, affordable, accessible,

Figure 1.3 The UN Sustainable Development Goals (SDG) (www.un.org/sustainablede-
velopment) (The content of this publication has not been approved by the
United Nations and does not reflect the views of the United Nations or its
officials or Member States)

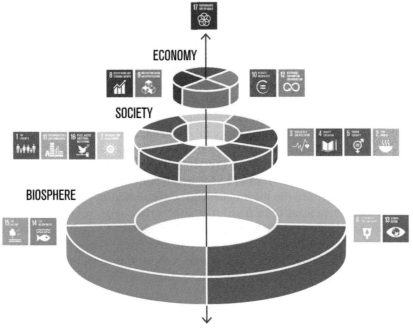

Figure 1.4 Example of an integration of the UN SDGs with triple bottom-line model of
sustainability (Bergman, Z. et al (2018)).

Figure 1.5 SDG goals supported by World Green Building Council (WGBC) (www. worldgbc.org).

and sustainable transport systems for all; strengthen efforts to protect and safeguard the world's cultural and natural heritage; and to reduce the adverse per capita environmental impact of cities. The targets are to be fulfilled by 2030.

This and other important SDG goals are supported by the WGBC, as shown in Figure 1.5. They are described in more detail below with comments from the WGBC:

- *Goal 3: Good health and well-being* – To ensure healthy lives and promote well-being for all people at all ages. The design of a building can influence the health and well-being of its occupants. Lung and respiratory diseases are associated with poor indoor environment quality. Building characteristics, such as improvements in lighting, air quality, greenery, and foliage, have gained interest in recent years because of their positive impact on health and well-being. Such improvements also reduce emissions from buildings, especially in urban areas.
- *Goal 7: Affordable and clean energy* – To ensure access to affordable, reliable, sustainable, and modern energy for everyone. The cheapest energy is energy not used, and energy savings from efficient green buildings, whether commercial buildings or homes, are the most beneficial. Green buildings that use renewable energy is more beneficial than fossil fuel alternatives. For example, home solar systems in Africa can provide households with electricity for as low as $56 a year, which is much less expensive than energy from diesel or kerosene. Renewable energy also has the benefit of producing zero carbon emissions, limiting its impact on climate. Energy efficiency linked with local renewable sources also improves energy security.
- *Goal 8: Decent work and economic growth* – To promote inclusive sustainable economic growth through employment and decent work circumstances for everyone. As the demand for green buildings grows globally, the workforce required for the delivery of these buildings has grown. Additionally, the life cycle of a green building, from conception to operation, maintenance, and even deconstruction,

provides more opportunities for inclusive employment. Some developing countries have developed ways to integrate more socio-economic issues, such as unemployment or lack of skills into their green building rating systems, creating further incentives for businesses to consider these criteria in their development.

- *Goal 9: Industry, innovation, and infrastructure* – To build resilient infrastructure, promote sustainable industrialisation, and promote innovation. Green buildings must be designed to ensure that they are resilient and adaptable to the changing global climate. This is very important in developing countries, many of which are particularly vulnerable to the impacts of climate change. However, it is not just about resilient green buildings but also the spaces in between, i.e. the infrastructure that must be equally sustainable and resilient to future risks. Huge investments must be made worldwide in the coming decades in the infrastructure sector to achieve successful net zero emissions in the future. The aim of buildings to push the boundaries on sustainability, such as net zero emissions and beyond zero emissions buildings, is also a major driver for innovation and technology.
- *Goal 12: Responsible consumption and production* – To ensure sustainable consumption and production patterns. This goal focuses on promoting resource and energy efficiency and sustainability in infrastructure by providing access to basic services and green jobs. The construction industry plays a major role in refining waste through reduction, reuse, and recycling, while circular economy principles focus on resource productivity.
- *Goal 13: Climate action* – Taking urgent action to fight climate change and its impacts. Buildings are responsible for over 30% of global greenhouse gas emissions and are a major contributor to climate change. However, green buildings have immense potential to contribute to the decrease in emissions, offering cost-effective ways through energy-efficiency measures.
- *Goal 15: Life on land* – Promoting sustainability management of forests, decreasing desertification, halting and reversing land degradation, and halting biodiversity loss. The materials used to build a building is key to determining its sustainability. The building industry and its supply chains play a major role in the use of responsibly sourced renewable materials, such as timber. Green building certification tools also recognise the need to reduce water use, considering the value of biodiversity and the importance of its protection. Incorporating this into building sites, both during and after the construction stage, can minimise damage and enhance biodiversity, for example, through landscaping with local flora.
- *Goal 17: Partnership for goals* – Adopting an overall objective for the revitalisation of a global partnership for sustainable development in construction. Historically, the construction industry lacked a collective voice on the world stage at major climate change conferences and has not been recognised for the immense opportunities the industry presents. However, a significant milestone was achieved in 2015 when the Global Alliance for Building and Construction was launched. With strong partnerships, the capacity to drive a change is increasing, with recognition that overcoming barriers to a sustainable built environment involves not only technical solutions but also effective industry collaborations, ensuring that collective efforts are aligned to achieve much greater impact.

A large proportion of the earth's resource turnover is linked to buildings and construction. This implies a complex relationship to consider in the realisation of construction projects since economic and social aspects also affect the individual. To gain a holistic perspective on this relationship, during the construction process (the construction process specifically, see the section about the process further on), one solution is to adapt the concept of sustainable development to the construction sector and to direct the construction process accordingly.

Construction contributes to sustainable development, with regard to the five capital model mentioned above, through man-made (built), human (labour force), social (human well-being), and environmental capital to the capital stock with technological (productivity, labour, and material productivity) change.

The construction sector is complex and fragmented, and resistant to changes leading to increased sustainability. During the transition to a zero carbon, resilient, and sustainable society, buildings play a significant role in the use of energy and carbon emissions in most countries. In industrialised countries, construction typically has an economic importance of approximately 10% of the country's GDP. Globally, buildings are estimated (2020) to consume approximately 35% of the total available energy (producing approximately 38% of total carbon emissions) and to generate approximately 36%–40% of all human-made waste. The issue of sustainability occurs at all levels in the economy connected to the construction sector: macro, meso, and micro. At the macro (global) level, the construction sector declines as the country develops and, so far, sustainability has become more important in industrialised countries with a declining share of construction. This implies that at the global level, the construction sector is a poor contributor to sustainability. In recent years, however, there has been a change as more developing countries have begun to address sustainability issues in construction. New buildings are a significant source of future emissions, especially in countries with rapid economic development and growing populations, where most of the doubling of floor space is expected to occur by 2050 (see the section about population increase and construction). However, in many middle- and high-income countries, the existing building stock will represent the majority of building floor area in 2050; thus, action must be taken to improve existing buildings through a sustainable and regenerative transition. At the meso-economic level, the construction sector is composed of products and services supplied by many other industrial sectors, so it is difficult to ensure sustainability along the entire supply chain, especially with emerging global markets and international trade. On the micro level, a single building tends to be erected over a shorter time span because the client or investor is facing a more uncertain economic environment; thus, it is best suited as a short-to-medium-term economic asset for the owners.

Building and planning are key issues in municipal and regional development. It is difficulty for construction and real estate companies to benefit financially in their construction projects when sustainability issues are addressed. At the same time, the demand is increasing for ecological sustainability and contributing to reduced exclusion through social sustainability regarding both new buildings and renovations. In terms of ecological and social sustainability, experience indicates that it is often difficult to implement and transfer the initial ambitions in construction projects to reach fruition. This is due to a lack of competence but also because sustainability criteria can be difficult to reconcile with short-term financial interests. Compared to the economic and ecological aspects of sustainability, social and cultural sustainability criteria are relatively new in construction. Architecture, building planning, service in the local environment, and work environments can be considered examples of social sustainability. Cultural heritage, diversity, and cultural expression in the construction industry are examples of factors linked to cultural sustainability.

The implementation of sustainability for construction depends to varying extents on the global, national, regional, local, corporate, and individual levels. It also depends on the cultural and social context of the society in question, and many industrial countries have already established national strategies for sustainable development to measure their national or regional share of global depletion of resources. For the construction sector, these national strategies imply policies for sustainable buildings and sustainability in construction. From the public's perspective, the current rendition of policies and objectives are directed towards a single client in the form of different kinds of incentives, such as taxation subsidies, direct investment subsidies, public procurement advantages, and allowance of specific investment funds. It could also be a matter of sector agreements about overall objectives and specific targets regarding sustainability matters.

Somehow, all these differences have the same meaning in the end because the concern is how current decisions are going to affect future well-being, that is, changes in real asset values. However, the impact on common construction projects and their stakeholders remains unclear. In addition, there often appears to be a lack of knowledge transfer to local construction project management regarding how to manage the construction process towards sustainability. The local cultural context is significant when challenging and reforming the normative expectations of both construction professionals and end users.

1.2 The climate factors

In this and the following chapters, the term carbon emission includes all greenhouse gases, such as carbon dioxide, methane, hydrofluorocarbons (HFCs), and nitric oxide. While the correct term is carbon dioxide equivalent emission, for the sake of readability, it will be referred to as simply carbon emission.

The Hothouse Earth

Human activity has impacted the climate by increasing the global average temperature to more than 1.2°C (2021) above the pre-industrial level. A brief compilation of the human impact on key factors of climate systems indicates that there is a huge risk of reaching the state of 'Hothouse Earth' once the temperature increases 2° above the pre-industrial global mean temperature. Hothouse Earth is a very extreme state when several global climate systems reach their tipping point, pass a threshold of irreversible change, and combine to make the earth much hotter and a much more hostile environment for life. It has been estimated that this extreme state will last for at least 1 million years before declining.

The concentration of greenhouse gases (carbon dioxide, methane, and nitric oxide) in the atmosphere has reached approximately 413 ppm (2020). According to the latest synthesis report by the Intergovernmental Panel on Climate Change (IPCC, 2021), it is highly likely that this represents the highest concentration of atmospheric greenhouse gases in at least 2 million years. The annual average increase in carbon emissions has almost quadrupled since 1960.

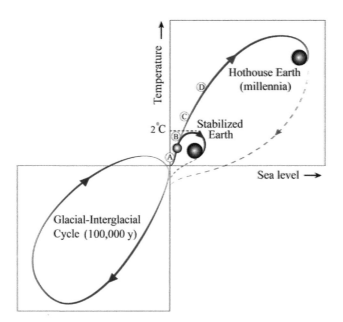

Figure 1.6 The Earth's system normal trajectory and to the Hothouse Earth status (Steffen, W. et al (2018)).

As shown in Figure 1.6, the normal status of the Earth's climate has been cyclic during the last couple of million years, from glacial (ice age) to interglacial status (a period of warmer climate), i.e. the glacial–interglacial cycle. This cycle lasted for approximately 100,000 years, as shown in Figure 1.6, where the horizontal axis refers to sea level and the vertical axis to temperature. The last period of warmer climate started approximately 11,000 years ago and is called the Holocene, which peaked about 6–7,000 years ago. The average warm period in the cycle lasts for approximately 10–15,000 years, after which the glacial period begins again. However, over the past 6,000 years, human activities, such as increasing farming and decreasing forests, have increased the carbon emission concentration in the atmosphere to a level that has caused the end of the warm period to be postponed.

Point A in Figure 1.6 illustrates this postponed status, which is quite stable with a variation of approximately plus/minus 1°. Point B illustrates the situation approximately 125,000 years ago when the average temperature was about 1°–1.5° above the pre-industrial level, but the carbon concentration was much lower than today (approximately 280–300 ppm). Points C and D on the path to Hothouse Earth illustrate the situation that occurred 2–3 million years ago compared to 15–17 million years ago when the atmospheric carbon concentration was approximately 400–450 ppm and 350–500 ppm, respectively. The average temperature was 2°–3° compared to 4°–5° above the pre-industrial level. The mean sea levels also changed during those times, from 6 to 9 meters above the current level at point B, to 10–22 meters at point C, and up to 60 meters at point D.

It seems that 2° (the dotted line in Figure 1.6) is the threshold for avoiding the Hothouse Earth path, which is also the limit established by the Paris Agreement

COP21. If the goal of 2° is met, it is very possible that the global status of point C occurs at a primary stage that will decline over time to the sea levels of a stabilised Earth, as shown in Figure 1.6. Then, after several 10,000 years, the Earth will return to its normal glacial–interglacial cycle.

The rapid increase in carbon emissions and the subsequent rise in temperature have triggered several global key climate systems to reach their tipping points towards an irreversible state. When one system collapses, it triggers the collapse of other systems in a sort of domino-effect, tipping cascade (see Figure 1.7).

The temperature has already risen 1.2° and caused increasing thaw of alpine glaciers, Greenland glaciers, seabound glaciers in the West Antarctic, and the loss of summer ice from seabound ice in the Arctic. This ice loss triggers changes in deep-sea currents when excessive cold and fresh water meets warm and salty water, which can trigger a faster thaw of ice sheets and glaciers. When the total area of reflecting snow and ice sheets decrease, the heat absorption increases in the sea and on the ground. Concurrently, the sea level rises when new fresh water is released from the melting ice sheets and water expansion occurs due to the warmer sea surface. It has been calculated that sea levels have risen approximately 5, 7, and 12 meters from melting of West Antarctic, Greenland, and Eastern Antarctic ice sheets, respectively. Another observed change regarding sea current is a collapse of the Atlantic Meridional Overturning Circulation, the immense Atlantic current that provides the Northern Atlantic with warm surface water and makes the climate milder on the east coast of North America and Northern Europe. This current has declined during the last decades and is very close to the tipping point of collapse which implies a transition to a weaker and slower mode of the current (the current has two distinct modes – strong

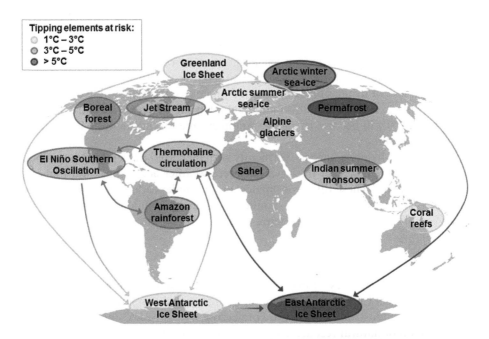

Figure 1.7 Important tipping points linked to a tipping cascade (Steffen, W. et al (2018)).

and weak), which would severely impact the global climate. Other tipping points that may occur around 2° above the pre-industrial level include the thaw of permafrost in the tundra of the northern hemisphere, causing a sizable emission of methane (a very strong greenhouse gas), an increase in bacterial respiration of carbon emissions in the sea due to rising sea temperatures, dieback, fires, and fast temperature changes in the Amazon and Boreal forests, thus, decreasing possibilities of absorbing carbon as carbon sinks.

The collapse of all of these systems impacts all of humanity's significant systems, including agriculture, fishing, biodiversity, fresh water supply from melting glacier ice, or lack of access to fresh water supply from groundwater, flooding of coastal areas and river deltas, disturbance of local monsoon regularity, and severe weather conditions.

For the status of a stabilised Earth in Figure 1.6 to become a reality, we must immediately reduce carbon emissions by managing different key activities to increase forest areas, bind more coal with new cultivation methods in agriculture, bind more coal by increasing biodiversity on ground and in water, and finally, develop artificial carbon sinks such as carbon capture storage. Furthermore, proactive efforts must be made to reduce the use of fossil fuels, emissions from cattle, rice cultivation methods, and the use of fertilisers in agriculture. It is also important to reduce the use of cement by using more renewable materials, such as wood, or by reusing materials in a recourse effective way via a circular economy. Consumer demand and technical innovations are also important ways of reducing carbon emissions.

This question of climate change is the most important issue for our and future generations' way of living. From now until 2030 is the most crucial period during which the increase in carbon emissions must turn to decline, through extreme global proactivity. Many changes in the climate system have become more extreme due to increasing global warming, including an increase in the frequency and intensity of heat extremes, marine heatwaves, heavy precipitation, agricultural and ecological droughts in some regions, the number of intense tropical cyclones, and reductions in Arctic Sea ice, snow cover, and permafrost. The densely populated coastal regions especially must withstand a mean sea level rise of up to 10 meters during this century. Two-thirds of the world's megacities are situated less than 10 meters above the sea and about 90% of urban areas are situated on coasts and vulnerable deltas. When a rise in mean sea level occurs, it is more likely that the frequency of extreme sea levels (see Figure 1.8) will become a more common event and, thus, a more frequent threat to coastal habitats.

Most human habitats have had difficulty with increased perception, more flooding, severe dry periods, lack of fresh water, severe weather conditions, and higher average temperatures. These different impacts are not independent of each other but are intricately interconnected so that significant differences could occur regionally, both within and between regions (see Figure 1.9).

Carbon Law

During the last few years, efforts have been made to describe how to reduce carbon emissions to meet COP21s target of maintaining a global temperature no more than 2° above the pre-industrial level or, more preferably, a limit of 1.5° above. The increasing consumption of natural resources by humanity is the main cause of this global environmental depletion and climate crisis. Further, there is a lack of understanding about

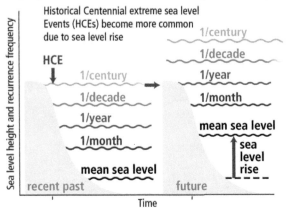

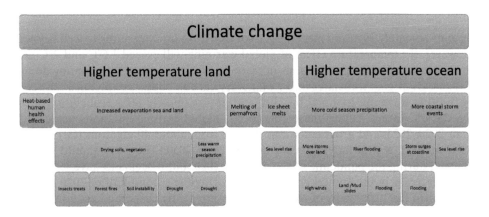

Figure 1.8 Regional effects of future sea level rise (IPCC (2019)).

Figure 1.9 Interconnected Climate change impacts (iiSBE (2016)).

the Earth's limited ability to assimilate residues from consumption, i.e. the Earth's resilience. To minimise this impact, risk management and sustainable solutions are essential to promoting recovery and alleviating the climate crisis.

One way of looking forward is to adapt a roadmap towards zero carbon, where the ambition is to cut all carbon emissions in half every decade from 2020 to 2050 and to establish a zero carbon emission status, as shown in Figure 1.10. This transition is necessary but also achievable. The challenge of zero carbon development in terms of a global multi-decade roadmap could be based on 'Carbon Law', which halves gross carbon emissions every decade. It can be complemented by immediately generated, scalable carbon removal from land use carbon emissions. It can lead to net zero emissions

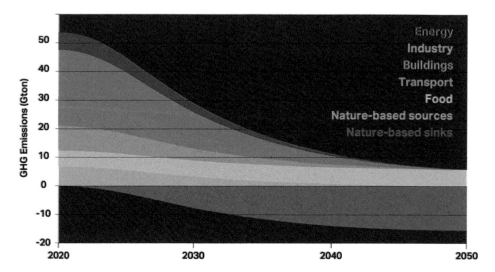

Figure 1.10 Carbon Law, the exponential roadmap to 2050 (Falk, J. et al (2020)).

by approximately 2050 and meet the goal of COP21 to limit warming to well below 2°C. However, the temperature effect of global warming will exceed 1.5°C and 2°C during this century unless deep reductions in carbon emissions occur in the coming decades.

The COP21 goal translates into a planetary carbon budget, implying a 50% chance of limiting warming to 1.5°C by 2100 and approximately 66% probability of meeting the 2°C target. Under these conditions, global carbon emissions peak no later than 2020–2023, and gross emissions decline from about 40 gigatons (Gt) in 2020 to 5 Gt in 2050, with a halving every decade 2020–2050. The peak should have been reached during the pandemic of Covid-19 in 2020, but it seems to have led only to a slight decrease in the annual growth rate. During the second half of 2020 and whole 2021, when the lockdowns of the pandemic were partly eased and economies began to recover, the carbon emission growth rate also seemed to increase towards annual pre-pandemic levels.

Following such a global carbon law means at least limiting cumulative total carbon emissions until the end of the century from about 700 to 200 Gt, which allows for a small but essential uncertainty regarding the risks of biosphere carbon feedback or a delay in carbon removal or carbon storage. Farming and other land uses must store carbon instead of emitting greenhouse gases. Large-scale reforestation and forest, wetland, and peatland management need to protect the resilience of vital Earth systems. Finally, robust technologies for storing carbon, e.g. Bio Energy Carbon Capture and Storage (BECCS), must be developed. However, these technologies are controversial and have never been used on a large scale because of the high cost and risks of unintended consequences. BECCS should be used as insurance and not as a primary solution. However, as the carbon budget is consumed, it will become increasingly necessary. The budget of carbon emissions per year must decrease from 4 Gt by 2020 to 2 Gt by 2030, to 1 Gt by 2040 and to 0.5 Gt by 2050.

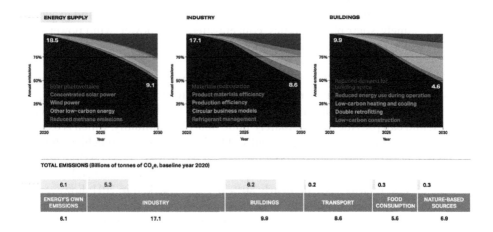

Figure 1.11 Carbon Law applied by sectors (Falk, J. et al (2020)).

Carbon Law applies to all sectors (see Figure 1.11), and countries of all sizes must immediately act in order to meet the 2030 requirements, to meet the first plan for halving carbon emissions. Roadmaps can be used as planning instruments to link shorter-term targets to longer-term COP21 SDG goals. They help align actors and organisations to instigate technological and institutional breakthroughs to meet a collective challenge. A carbon roadmap for halving carbon emissions every decade, designed by and for all industry sectors, could help promote technological advances towards a zero carbon world.

The sectors of most emissions include energy, transforming energy, industry, and buildings. Together, they emitted approximately 33 of 54 Gt (approximately 64%) on an annual basis by 2020 (see Figure 1.11). The energy sector itself emits approximately 6.1 Gt related mostly to extraction, refineries, and distribution of fossil fuels. The energy sector is also responsible for emissions from other sectors' emissions from their energy consumption, which totals about 12.3 Gt.

Energy sector

After being relatively constant, the market share of fossil fuels in the energy sector (and, therefore, the carbon intensity) has decreased over the last few years owing to the growth of renewable energy. Solutions to halve emissions by 2030 are available, and the changeover has already begun. Improved energy efficiency in all sectors (see below per buildings and industry) is economically important. Halving of emissions generated directly by the energy sector consists of emissions from extraction, refineries, and distribution of fossil fuels (6 Gt) and the generation of electricity and heat (12.5 Gt). To achieve a 50% decrease in its carbon emissions by 2030, the global primary energy demand should be reduced by efficiency gains in energy conversion, e.g. industry and buildings. This assumes that electricity consumption will level out from 2020 to 2030 as efficiency gains and savings elsewhere equalise an increasing number of end users and share of electrification. Nearly, a billion people lack access to electricity (2021) and

3.3 million people die prematurely annually from outdoor air pollution, often due to the burning of fossil fuels. Five key solutions can play integral roles and grow quickly enough – namely, solar photovoltaics, concentrated solar power, wind power, reduced methane leakage, and grid flexibility and storage.

- *Solar photovoltaics (PV)* – Solar PV panels convert sunlight into electricity and supply local or regional grids. They can be placed on buildings or the ground. The solar PV technique can provide more than all the current energy used globally, and can provide approximately 1,000,000 TWh annually, which is 6,000 times the current total energy used globally. The PV technique can achieve both exponential cost reduction and exponential capacity growth. Its growth is not limited by the construction of power plants, as factories produce solar cells and panels. The highest relative growth, nearly 80% in a year, peaked in 2011; however, global growth has slowed slightly to 20%–25% per year since then. Contributing to the 2030 goals for the energy sector, solar energy should continue growing exponentially at a rate of 20%–25% per year, reaching 6,000 TWh/yr. and reducing emissions in the sector by 4 Gt/yr. in 2030. This rate represents less than half of the highest historical growth; that is, solar energy could potentially drive emissions down even faster.
- *Concentrated solar power (CSP)* – Concentrated solar power derives from a large appliance that produces solar energy on a large scale. It involves mirror systems that are used to concentrate sunlight on a small receiver. The heat from the receiver then drives the turbine to generate electricity. Unlike PV, CSP requires direct sunlight to collect energy and is mainly situated in arid regions. However, the collected energy can be stored as heat and used to produce power at night. CSP plants can balance daily fluctuations in the energy system and complement the use of solar PV. CSP requires frequent cleaning as dust can seriously impact performance. The power output is more sensitive to solar radiation which may change depending on the weather and solar angle, and an accurate tracking system for focusing solar power is required. CSP could contribute to a significant share of renewable energy, and its relatively low-cost storage capability provides a competitive advantage. The growth could be an exponential of 40% per year until 2030. This would contribute to a reduction in carbon emissions in the sector by 0.4 Gt/yr. in 2030.
- *Wind power* – Approximately, 2% of the solar energy striking the Earth's surface is converted to wind energy. Wind energy is then converted into electricity by wind-driven turbines. It is still a larger source of electric power than solar PV and low-cost technology per unit of energy. Wind can be produced on land (onshore) or at sea (offshore), and when connected to the grid, wind power can be produced without subsidies. Several countries have successfully reduced their carbon emissions by changing from fossil-based power systems to wind power. Wind power has grown exponentially since 2000. By the end of 2018, nearly 600 GW of wind power was installed with the capacity to generate 5% of global electricity, 1,300 TWh/yr. Future technical potential for wind energy, such as solar techniques, could exceed current global energy use; for example, commercial wind turbines with a mean height of 88 meters could potentially provide 840,000 TWh/yr. globally, and the global demand is approximately 25,000 TWh/yr. With a reduction in energy supply emissions by half by 2030, wind power should continue to grow

at a rate of approximately 10% per year, reaching 4,000 TWh/yr. and reducing emissions by 2.0 Gt. As with solar PV, its recent growth has been slower than past rates. Nevertheless, future development of wind power technology could cut emissions even faster.

- *Other low-carbon energy* – New nuclear generation, hydro power, wave power, geothermal power, and heat/power cogeneration capacity with biomass fuel will also contribute to emission reductions in energy supply. With these alternatives, the combined contributions by 2030 are estimated to be 0.9–0.95 Gt per year, with none of them contributing more than 0.22 Gt each.

- *Grid flexibility and storage* – The energy sector does not use digital technology in the same way as other sectors. Often, energy systems are operated as local monopolies in a conservative manner, rather than being open to innovation. However, using digital technologies offers great potential to cut emissions. Managing the power distribution digitally and storing it by battery can provide stability. Pricing in real-time and web-controlled applications could give customers the ability to reduce their cost by flexible demand, resulting in a more stabilised grid. With the high-level use of low-cost renewables, efficient transmission of power will also become more important when supplying large cities and industrial regions with high demand from areas where renewables are produced. Because electricity from renewable sources is available at a lower cost than from fossil fuels, electricity could also be used as so-called 'electro fuels'. These are a kind of energy storage or flexible demand that could make an impact by 2030, but the scale is currently difficult to quantify. With increasing production from intermittent renewable sources and the need to balance the grid, batteries are a key component of a low carbon energy system. Batteries are already used to balance grids at low cost, while also making the growth of 24/7 off-grid solar electricity possible. A key is the cost, but batteries are becoming cheaper to produce as demand from other industries has increased the scale of production. Estimations indicate that global battery capacity will grow exponentially, at a rate of doubling every 28 months until 2030. Although uncertain, the estimation is that battery storage together with technologies for flexible and optimised grids can decrease fossil-based electricity by 10% by 2030. This reduces emissions by 1.6 Gt/yr. by 2030.

- *Reduced methane leakage* – Methane is a strong greenhouse gas, which is approximately 30 times stronger than carbon dioxide. It is emitted when leakage occurs during fossil fuel extraction and when fossil methane gas leaks during transport in pipes or as liquefied natural gas on ships. An estimate of methane emissions from fossil fuel extraction and distribution is approximately 2.7 Gt. Solutions to reduce this leakage are available, but they are not applied at a large scale because the payback is considered too low. A stronger policy and better monitoring techniques can help close the gap. Available technologies include optical and portable tools to measure and drones to detect and monitor leaks. Reducing the leakage of methane is often both profitable and environmentally beneficial. With stricter policies, cooperation in the industry sector, and application of the latest technology, emissions of methane from fossil fuels could be reduced by 0.45 Gt by 2030, with the possibility of a net economic gain. The full technical potential was almost three times higher. The reduction of methane leakage should be considered a short-term solution to rapidly reduce fossil fuel emissions, while the main strategy is to replace all fossil fuels. Methane leakage from agriculture and waste in landfills can also be used to replace fossil fuels. Such opportunities could be significant globally.

Buildings

Annual carbon emissions related to existing buildings are approximately 9.7 Gt, with about 60% coming from residential and 40% from non-residential buildings, respectively. Annual emissions related to building construction, on the other hand, are increasing steadily and have reached approximately 3.7 Gt.

Humans are using buildings as shelters all over the world every day, but until quite recently, their impact on the climate had not been sufficiently considered. According to carbon law, emissions in this sector will halve by 2030, and the use of buildings will also change in both positive and negative ways. On the one hand, new technology is helping to lower energy use through better materials and digitalisation of operation. However, on the other hand, the total area of buildings is growing rapidly globally, resulting in higher emissions from the construction process and from usage. The predicted increase in the building area is approximately 200 billion square meters by 2050.

The common way of measuring energy use in buildings is kilowatt hours per square meter (kWh/sqm). To determine a building's lifecycle perspective energy use, the total energy use of a building is divided by the expected lifetime of the building. Further, the benefits of the building are another aspect to be considered. The common key figure of energy use should be complemented with a part of the intensity of use – for example, energy use per resident or client (kWh/sqm and person), depending on the purpose of the building.

There are several methods for reducing carbon emissions and energy use. One possible method is to reduce the demand for building space; another is to reduce energy use during operation; another is to use low carbon, heating, and cooling systems and, of course, to minimise the use of energy and carbon emissions during construction, refurbishment, and retrofitting, see Figure 1.12:

- *Reduce building space* – Some sectors have reduced their need for building space considerably in recent years (mostly banks and post offices), by moving most of their onsite activities online. In the coming years, possible changes in the demand for building space could be in education, due to increasing online education, especially as a result of the Covid-19 pandemic which has caused schools to use online methods. Moreover, also affected by the pandemic, different kinds of retail stores have increased online shopping in lieu of onsite space demand. By more efficient use, building space can also be reduced. This can be achieved by more intense use per square meter or per day; for example, flexible seating in offices increases the space usage, or using a school building for non-school activities in the evenings increases the level of use per day. Residential buildings can be used more efficiently by increasing the number of residents per floor area or by increasing the use of home offices. Shared workspace solutions can provide more than twice the space efficiency compared to traditional office solutions.
- *Reduce operational energy use* – The operational energy used by a building can be reduced by adapting temperature, ventilation, and lighting in accordance with usage through digital sensors or by artificial intelligence (AI) systems, which can

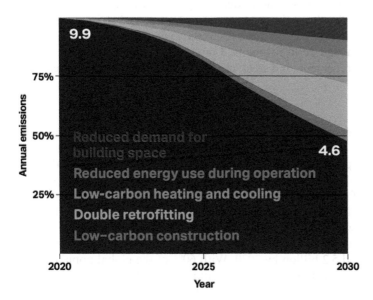

Figure 1.12 Carbon Law applied by the building sector (Falk, J. et al (2020)).

'learn' the usage of a building and predict changes in advance. An AI system can interact with a smart grid – for example, it can heat a building when the energy supply is high, and demand is low. This would allow the building stock to balance supply variations in an energy system with a high share of renewables.

- *Low-carbon heating and cooling* – Heating and cooling in buildings contribute to high carbon emissions. It can be reduced with low-carbon systems such as heat pumps, solar cells, heat storage systems, and district heating and cooling systems based on renewable resources. Rapid transformation to renewable energy production, especially local production, and heat storage at the building and/or block level offers a key opportunity to decrease carbon emissions.
- *Doubled retrofitting* – Doubling the speed of retrofitting globally among existing buildings offers a huge potential for reducing carbon emissions (see Figure 1.20). Improved insulation, LED lighting, energy-efficient ventilation, and smart windows are the most important measures with high impact. It is also important, from a lifecycle point of view, to consider a circular and a low carbon system regarding the production of new insulation, ventilation, and windows; otherwise, planned carbon emissions savings can be lost.
- *Low-carbon construction* – Carbon emissions from the construction process and the production of construction materials, such as cement and steel, are substantial and must be specifically addressed. The lowering of construction emissions has not been in focus until recently, and efforts to minimise these emissions in calculations based on LCA are increasing globally. The reuse of building structures is a priority to avoid new carbon emissions if possible, but if new buildings must be constructed with unused materials, low-carbon products should be chosen. In addition, it is also very important to optimise material usage space, and energy-efficient designs, as well as locally sourced materials and low-carbon transport.

Shared construction process knowledge, usage of building information model (BIM) technology for the entire construction process, and stricter sustainability demands in the procurement processes, should strongly reduce carbon emissions from construction and refurbishment. Emissions from construction materials, especially concrete, cement, and steel, are described in greater detail below.

Industry sector

The industry sector is responsible for 17 Gt of carbon emissions annually, representing 32% of the global total. This figure can be divided into two parts. First, energy-intensive heavy industries that produce materials – such as steel, cement, plastics, aluminium, and chemicals – create substantial emissions. Second, light, less energy-intensive industries, including fashion, furniture, and home appliances.

Carbon emissions from heavy industries have grown exponentially for several decades. Further growth has been predicted, unless measures to reverse the growth are applied. The demand for industrial goods will increase as the global middle class increases to a projected 5.2 billion by 2030. With 60% of the global population living in urban areas by 2030, growing cities will demand more buildings and building materials. Without rapid transformation to a low-carbon or zero carbon sector, building materials such as concrete (including cement), steel, and plastics will easily consume the 1.5°C carbon emission budget.

However, several large industrial companies have set climate targets and are implementing circular production models. Low-carbon materials are becoming more common in all types of products. Industries realise that incorporating strategies to include the Paris Agreement, COP21, offer a competitive advantage. The key solutions are to reduce carbon emissions by a few key principles, such as implementing a circular production process, making products with less quantity of materials, reusing materials, replacing high-carbon materials with low-carbon materials, and optimising production processes.

By reducing the quantity of materials used in steel, concrete, cement, aluminium, and chemicals, or by increasing their usable lifetime (material productivity), costs can be lowered as emissions are reduced. Additionally, the reuse and recycling of materials is a key opportunity to reduce carbon emissions. The quantity of recyclable material globally is steadily growing, and energy savings are often in the range of 60%–75% for the usage of recycled materials. Corresponding savings for steel are up to 90%. Achieving these savings requires products to be designed for disassembly and recycling, avoiding material contamination and improving recycled material collection rates and processes.

Construction projects can use 30%–50% more steel and cement than necessary and increasing material productivity through the entire life cycle of the product is an opportunity to reduce both carbon emissions and costs. Large reductions in carbon emissions are also possible by substituting high-emission materials with equal or better low-emission materials, e.g. low-carbon cement. Improved codes and requirements backed by BIM design technology will be necessary.

A large part of carbon emission reduction for steel, cement, plastics, aluminium, and other materials is to improve production efficiency. The energy intensity could be reduced by up to 25% by 2030 by upgrading or replacing existing equipment with the best available technology. Using renewables in the production of electricity and heating/cooling will provide further cuts. Additional cuts of up to 20% of the annual energy intensity could be possible by measuring processes and energy use in real time, using AI techniques for optimisation.

Adopting a circular economy model (see Chapter 2) in operations and supply chains is becoming a business advantage. A circular economy model is used to maximise the circulation of products, components, and materials through reuse and recycling, and the value bound to them as much as possible in the economy. This creates real economic and social benefits. A more circular economy could cut carbon emissions from heavy industry, e.g. 45% of carbon emissions from steel, cement, plastic, and aluminium products by 2050 globally. Additionally, another huge opportunity to reduce emissions simultaneously is business models, where buildings, tools, and vehicles, which could be vacant for up to 90% of the time, are released for others to use.

When depletion of the ozone layer was discovered in 1984, UN member states agreed to phase out the main cause – refrigerants of chlorofluorocarbon. The industry shifted to another refrigerant, HFC, which caused less damage. However, HFCs are also powerful greenhouse gases with an impact more than 1,000 times higher than carbon dioxide and have a long lifespan in the atmosphere. In 2016, nations agreed to phase out HFCs starting in 2019. More than 90% of the climate change impacts of HFCs can be avoided if emissions stop by 2030. The treaty, ratified by 65 countries, is projected to reduce global warming by 0.4°C this century.

However, the demand for cooling devices is increasing, especially in developing countries. Global warming with warmer air temperatures also drives the increasing demand for cooling. Low levels of device efficiency and high leakage of refrigerants drastically increase the impact of greenhouse emissions. Adopting a strong green cooling movement among suppliers and customers together with stronger regulated quality requirements of the devices could rapidly reduce emissions by, for example, using new types of refrigerants and better cooling efficiency.

1.3 Population growth and construction demand

In the big picture of the Earth's development, while life first emerged on the planet approximately 4 billion years ago, our species evolved into Homo sapiens in southern Africa approximately 2 million years ago (Figure 1.13).

Previous to human existence on Earth, there were a few mass extinctions, snow-ball earths, the development of photosynthesis to an oxygen-rich atmosphere, and the development of a protective ozone layer. We evolved alongside at least five Homo species during a challenging time, including a couple of glacial and interglacial cycles (see the section on climate factors). The cold period lasted 100–150,000 years and the intermediate warmer part lasted approximately 10,000 years. During this time, the mean temperature fluctuated by approximately 10 degrees up and down within decades. Approximately 75,000 years ago, the last cold period peaked and there was a significant shortage of fresh water due to the high level of ice sheets and the sea level approximately 70 m below today. Our species was close to extinction due to this water shortage; less than 10,000 fertile individuals survived in the highlands of today's Ethiopia.

The last ice age ended approximately 12,000 years ago, after which a warm period of stabilised climate occurred with a temperature variation of only plus/minus 1°C,

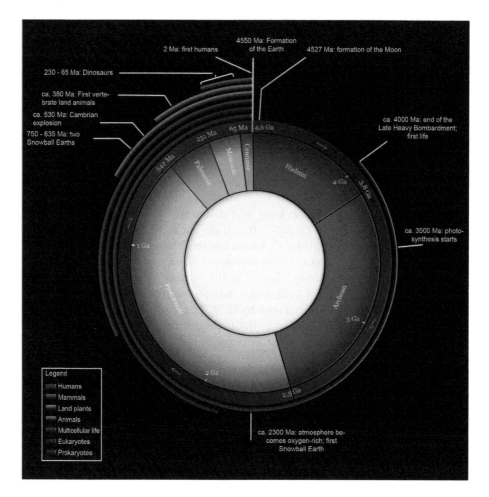

Figure 1.13 The geologic clock (https://commons.wikimedia.org/wiki/File:Geologic_clock.
jpg#filelinks).

a period geologists call the Holocene. Predictions have suggested that this relative stability last for a further 50,000 years. With such stability, the beginning of the Holocene promoted agriculture, and humans changed from hunters and gathers to farmers within a couple of generations. These farming societies have occurred simultaneously in different places on Earth. With farm-based societies, the human population emerged quite slowly from a couple of million individuals at the beginning of the Holocene to about 200 million around year 0, that is, approximately 100 times in 10,000 years. It took 1,800 years to reach the first billion around 1800, at the preindustrial level. As industrial development continued, the second billion was reached around 1930, the third in 1960, the fourth in 1974, the fifth in 1986–1987, the sixth in 1998–1999, and the seventh in 2011 (see Figure 1.14). Population growth has increased exponentially since the beginning of the industrial era. Industrial development began to mark a geological footprint around 1950 as the nuclear race intensified, and the first sign geologically made by humans was called the Anthropocene. Since then, the ceiling of earth's resilience was reached at the end of the 1980s due to the resource consumption caused by humans. With an increased affluence, higher level of education for all, decreasing size of the family, and a more equal allocation of resources, the rate of population growth predicts a decline over time. However, two-thirds of the projected growth through 2050 will also be driven by current age structures. It would occur even if births in high-fertility countries today were to fall immediately to around two births per woman over a lifetime. Globally, the generation of young people now entering their reproductive years is larger than their parents' generation; thus, even if the global level of fertility was to fall immediately to around two births per woman, the number of births would still exceed the number of deaths for several decades, and the world's population will continue to grow.

The UN prediction of growth until 2100 indicates that it will likely reach 11 billion, and the rate of growth will decline until 2060–2070, see Figure 1.14. Until then, the eighth billion will occur around 2023–2024, the ninth around 2040, and the tenth billion just after 2050. Other predictions include women's education level and increased birth control, and the 2100 level is set to about 9 billion, with a peak in the middle of 2060 of about 10 billion, a prediction of population decline for the last 30 years of this century.

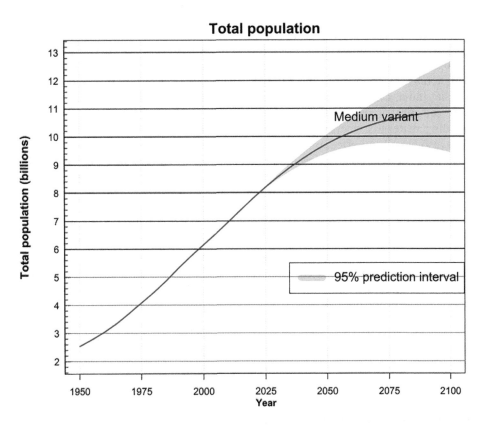

Figure 1.14 UN prediction of population growth until 2100 (www.population.un.org).

Until 2050, some 9–10 billion people require a place to live in residences or shelters, that is, during the next couple of decades, the world population will increase with at least one more China and India together. Half of this growth occurs only in nine countries, including India, Nigeria, Pakistan, Democratic Republic of Congo, Ethiopia, Tanzania, Indonesia, Egypt, and the United States. The largest country by population in 2050 and 2100 will be India, followed by China, Nigeria, the United States, and Pakistan.

More than half of the world's population is urban, with a continuing process of urbanisation from poorer rural areas and minor cities. Moreover, principally all the coming population growth will be in urban areas, that is, there will be more inhabitants in cities by the year 2050 than in the entire world in the year 2000. Currently, there are approximately 4 billion inhabitants in urban areas; in 2050, it will be approximately 6 billion, and most will be living in megacities with 20–30 million inhabitants each.

The demand for new buildings

Half of the world's population is predicted to have a higher mean income per capita by 2050 than the present mean level of OECD countries. It is predicted to increase by a factor of 13 in Brazil, Russia, India, China, and South Africa, and 50 times in the rest of the world. Together with population growth and mean income increase, the demand for new residential and commercial buildings with accurate quality and standards is very high. Demand also includes adaption to new and better social circumstances. This implies fewer persons per home and more floor space per person – to approximately 30 square meters per person in 2050. In total, the need for housing will almost double by 2050 in comparison to 2010, and non-residential service buildings will increase by about 70%, and half of the need will occur in BRICS countries (Brazil, Russia, India, China, and South Africa). To meet this immense demand for new buildings, the construction sector's need for resources will increase to a large demand for building materials, energy, and emissions produced in the manufacturing process. This also has a large effect on carbon emissions. The total budget of carbon emissions until 2050, to reach the 2° target, is to be reached by the construction sector alone if nothing is done to reduce its own emissions. The construction sector must reduce its own consumption of energy and carbon emissions to approximately one-fifth of the remaining carbon budget.

Buildings play a vital role in human society, including the environment, working life, and interaction between people. In the transition to a low-carbon, resilient, and sustainable society, buildings play a dominant role in the use of energy and are among the largest sources of carbon emissions in most countries. New buildings are an important source of future emissions, especially in rapidly economically developing countries with growing populations, where most of the expected increase in floor space is expected by 2050. However, in many middle-and high-income countries, existing building stock will support building of the floor area in 2050, also taking action to improve existing buildings critical to a sustainable transition.

Building construction and operations in 2019 accounted for the largest share of global total final energy consumption (35%) and energy-related carbon emissions (38%). Electricity consumption in buildings represents approximately 55% of global electricity consumption. Across the globe, building energy use constitutes a significant proportion of the overall energy demand. In 2019, buildings accounted for 57%

of the total final energy consumption in Africa and 32% of the total process-related carbon emissions. In ASEAN, China, and India, energy consumption in buildings accounted for 26% of the total final energy consumption and 24% of the total process and energy-related carbon emissions. Buildings account for 24% of the total final energy consumption in Central and South America, and 21% of total process-related carbon emissions.

The increase in building sector emissions is due to the continued use of coal, oil, and natural gas for heating and cooking, combined with higher activity levels in regions where electricity remains carbon-intensive, resulting in a steady level of direct emissions and growing indirect emissions, especially from electricity. Even though electricity has low direct emissions, it is still primarily sourced from fossil fuels, such as coal and natural gas.

Emissions from building construction and building materials are largely driven by cement and steel manufacturing, and their growth in use is a major driver of building-related embodied carbon emissions. Building design and type, such as high-rise towers, have resulted in an increased demand for steel and cement although such buildings may have a longer lifespan. Globally, the building construction sector accounts for approximately 50% of the demand for cement and 30% for steel. These factors show the importance of extending the lifetime of buildings, reducing cement and steel use, and replacing them with materials that have lower embodied carbon.

It is not yet clear what the impact of the Covid-19 pandemic will be on global energy demand and carbon emissions in buildings, but it has been estimated that overall global energy demand will drop by 5% in 2020 and energy-related carbon emissions will drop by 7%, which is largely due to industrial and transport-related changes in demand. This would be among the largest reductions in energy and carbon emissions observed in the past 30 years. The long-term implications of the global pandemic in the building sector seem, thus, to be marginal. While energy demand in buildings may be less impacted than the overall global energy demand due to the nature of building operations and lock-in effects of existing buildings, there may well be more drastic impacts on building design and use as well as integration with energy systems and other infrastructures. Covid-19's disruption in the buildings and construction sector seems to be of a minor importance.

1.4 The construction process and sustainability

The construction process is usually fragmented and complex with different types of clients who have different demands and requirements. There are different stakeholders throughout the process, all with different needs and interpretations. The products, i.e. the buildings, are situated at different places and sites, with different kinds of neighbourhoods, with different building cultures and aesthetic traditions. Depending on the region, there are different construction codes, different cultures of procurement, and, lastly, different regional and local climate conditions. The construction process itself, in the linear form, is divided into three main parts: the design, construction, and operational stages. The design stage is divided into pre-design – including inception, feasibility study, briefing, and conceptual design – and in design – which includes schematic design, design developments, and construction documents. The construction stage includes the tender phase and construction work. The operational stage includes operation, maintenance, refurbishment, and deconstruction or demolition. This is a classical linear description of the process. If using a timeline on this linear process, the first two are a fraction regarding time consumption to the third stage, that is, the design stage of a standard office building lasts for less than a year, the construction stage lasts about a year or two, and the operational stage could last more than 100 years. This means that the very short timeframe of design and the decisions made during this stage have immense consequences throughout the lifetime of the actual building. Furthermore, the resources used during the design stage, especially in the early phases, are a fraction of the total investment cost of the construction and a fraction of a fraction when considering the lifecycle cost of the building. Accordingly, if the design team considers the long-term implications of the chosen items of a construction project during a couple of more hours than the ordinary designated time, it could create immense economic, environmental, and social gains for the client in the long run. Another way of describing the construction process is the circular view, as shown in Figure 1.15, where the linear process is converted into a continuous circular process during the lifetime of the building with one endpoint, the end of life, the deconstruction.

The complexity of the construction process and its fragmented nature tend to resist changes towards a more innovative and changeable industry involving issues of balancing economy and social development with ecological considerations, i.e. the triple bottom-line. Some of the issues of main resistance are lack of proactive measures, conflicts regarding real and perceived costs (investment and lifecycle cost), and insufficient implementation expertise. Construction is an assignment with several different actors, each having different interests in dealing with multiple activities during a specific timeframe requiring the right level of quality to a specific cost on a given site. The possibility of changing the outcome of the product (i.e. the building) during the construction process decreases considerably with a higher level of design details and time. The work is undertaken by a project-based organisation, that is, the work starts from zero on a new site with new combinations of performers and, although, having

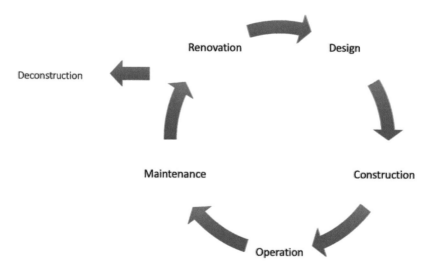

Figure 1.15 The circular view of the construction process (original by the author).

a common goal, the building. The uniqueness of construction works can be of two types – the product and the site where the building is situated. The site or location utilised by a building is a key variable in design and management decisions. When the building addresses aspects such as energy use, indoor climate, and material productivity, the site of a building addresses aspects at both the local and regional levels. Examples of local aspects are urban microclimates, accessibility to neighbourhood buildings, security, and local biodiversity. At the regional level, there are examples of aspects such as community demands, transportation systems, air quality, public health, and emergency preparedness.

Sustainability in construction

Sustainability in construction can be interpreted in several ways. The term covers a broad and complex interaction between stakeholders involved, aesthetics, functionality, and material interactions. Construction itself implies everything from site-specific activities to the creation of human settlements. Sustainability, on the other hand, should imply a holistic view – 'the whole is more than the sum of its parts' with relationships and interactions between humans, society, the biosphere, economy, and the state of technology. In this complex framework, it is the client/owner/developer, as the responsible performer of activities who has the main responsibility for construction projects and the obligation to commit sustainability. The client and the management team of a construction project with the aim of sustainability should, therefore, consider the entire process from the early design stage to the final product, and the benefits and negative impacts regarding the triple bottom-line of sustainability that are to be expected during the lifetime of the final product. With regard to the social aspects, it appears that information between stakeholders in a construction project is an essential missing part, especially regarding complex relationships and interactions related to sustainability issues. This means that it is important for the project management team to clearly and openly evaluate all possible options to obtain the project purpose with respect to the relevant issues of sustainability from the perspective of all project stakeholders.

As mentioned, the mainstream construction sector has been strongly focused on environmental considerations. The connection between sustainability and buildings as products is that regional culture and sustainability are complementary components, and the existing building stock is an essential part of regional and cultural diversity. It is also because of the value-loaded phrase of sustainability that different persons define the term differently according to their own view of society and how their view is accepted by others. Common terms used in the construction sector related to sustainability include sustainable building, sustainable construction, and green building. These are often interpreted differently by stakeholders in a construction project, depending on their education, age, cultural background, etc. Green building was the first term for a building designated more or less with special criteria regarding environmental issues, especially criteria covering a more efficient use of a building's energy performance. A similar expression is a sustainable building with a broader and more holistic focus on the triple bottom-line. These terms focus on the product, i.e. the building, and its outcomes and performance. Sustainability in construction is a broader expression that considers the entire process of construction from the early design phase to deconstruction, including a holistic view of the triple bottom-line and minimising the impact towards a net zero performance. When considering sustainability in construction, it must also be complemented with the specific conditions of the actual site, the ability or knowledge of the design and management team, and the local and regional conditions of the triple bottom-line. As shown in Figure 1.16, sustainability in construction can be divided into three dimensions: the object with varieties as region, urban, neighbourhood, building system, components, and material. The next dimension includes the process from early design through design, construction, operation, maintenance, and deconstruction; the third dimension, performance quality, includes the triple bottom-line complemented with functionality.

Currently, there are numerous examples of good practices regarding sustainability in mainstream construction.

One major measure to handle and to be more proactive with the client regarding sustainability is in the tender phase of a project, both regarding the design team and the contractor. Stringent and thoroughly made sustainability conditions in the tender phase are the client's opportunity to fulfil the obligation of sustainability in construction works as the responsible actor. The possibility of changing the outcome without significant additional costs during the rest of the construction process decreases considerably with time.

Certification of a building's energy and/or sustainability performance highlights best practice construction efforts and provides a market signal to investors, tenants, policymakers, or consumers with awareness of sustainability matters. Green or sustainable building certifications play an important role for clients and owners to separate their buildings within the market but also highlight the commitments to the principles of sustainability in building construction and operation. The use of certifications offers an assessment of how well a building meets the defined certification criteria and that relevant requirements are fulfilled. The certification of buildings and districts has

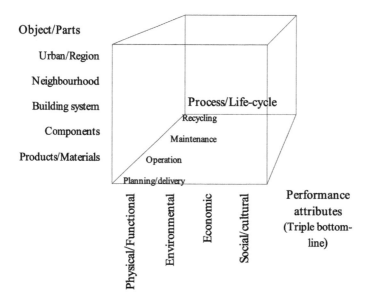

Figure 1.16 Sustainability in construction (Persson, U. (2009)).

also become an important tool for quality assurance during design, construction, and operation. Due to the lack of holistic and ambitious legislation in nearly all countries, and because most of the certification systems include the UN SDG-compatible levels of emissions, certification can fill the gap and work as a transformation tool of the market and its stakeholders to a building practice with more focus on sustainability. With increasing demand from the financial sector, certifications can also be used to verify the requirements of sustainable finance. In some low- and middle-income countries that lack or have outdated building codes, building certification could be used as a de facto building code and can be an important tool for securing both financing and tenants. For more about certification of buildings and the construction process, see Chapter 7.

1.5 Zero carbon and regenerative building

Buildings are functionally and culturally necessary for humans. Functionally, they provide a space for shelter and an environmental context for activities, providing a place of safety, sufficient space, and comfortable conditions. Culturally, buildings express the ambitions and technological capabilities of society. Buildings are complex, manmade creations. The design involves the interaction and coordination of a wide range of professions within a multitude of regulations and typically includes time and cost constraints. Construction requires the involvement of numerous skills and trades, and changes in operation and use can transform the initial design intentions. Buildings also represent a notable capital investment, both as financial and natural assets. However, due to more rapid societal changes and the long life of buildings, the replacement of buildings has historically been quite slow, rather than a fast momentary change. Consequently, renewal in the built environment is more determined by disasters, technological advances, revised regulations, and the costs/benefits of alternative strategies.

Technology development impacts building design. While air conditioning, electric lighting, elevators, and other building technologies have provided the user with greater comfort, the use of these components has significantly increased building operational energy and carbon emissions. Since the 1960s, environmental developments and advances have influenced social behaviour and environmental issues. This has slowly changed approaches to building design and construction. Figure 1.17 schematically positions some of the key societal concerns and the concepts and language used in the context of building design in response to them over the past 50 years. While they are presented chronologically for convenience, many of them overlap with each other, coexist, or reappear in different forms and have many interpretations. In the 1970s and the 1980s, strategies such as 'low energy', 'passive solar', and 'indoor air quality' characterised specific performances of buildings before they came under the umbrella of 'green' buildings in the 1990s. Green buildings specified a building's performance as doing less harm to the environmental impact, mostly by some sort of energy efficiency, as described above (see Figure 1.17).

During the first decade of the 2000s, this green approach moved its focus slowly from solely environmental issues and assessment methods regarding questions about energy savings and material productivity to a holistic view of sustainable building, including the triple bottom-line, to 'maintain what we have'. Since the 2010s, concerns about the impact of climate change on the building environment have increasingly become an issue. A zero carbon performance began to be emphasised together with a

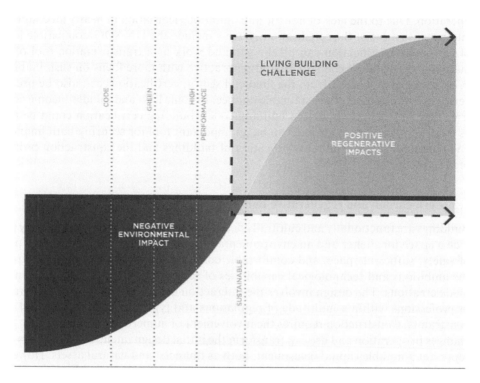

Figure 1.17 Past and future changes in building design concepts (Living Future Org (2021)).

shift from solely the performance of the product – the building – to the construction process, and the whole lifecycle perspective. From the beginning of the 2020s, it has been about zero carbon, zero energy, and a regenerative development including triple bottom-line, to 'retrieve what we lost' or 'doing more good', a net positive view, as shown in Figure 1.17.

Net zero

As a part of the global net-zero strategy, net-zero energy and zero carbon buildings must become the primary form of building construction across all economies by 2050. This is the most important issue to address for the global construction sector to meet and stay below the 2°C target mentioned previously. However, the concepts of net-zero energy or zero carbon emission buildings do not have a recognised standard definition. Most of the definitions have common elements such as very low energy consumption, buildings achieving zero energy or zero emissions during a year, when any energy consumed, or carbon emitted, is compensated by using renewable sources at the site of the building. Electricity source certifications are also suitable for reducing carbon emissions as power production shifts from fossil fuels to renewables.

It is possible to decarbonise the construction sector, but this requires the development and implementation of well-defined pathways for the crucial transition to achieve net zero carbon by 2050. There are many ways to support this transition, such as progressive building code implementation, market regulation, and support energy efficiency investments for existing buildings. Carbon emission reductions from buildings must be extensive and take place quickly, with substantial reductions by 2030 and almost complete decarbonisation by 2040. Direct carbon emissions should be reduced by at least 45% by 2030, 65% by 2040, and 75% by 2050 relative to 2020. The decrease in indirect carbon emissions from the power sector should occur more quickly. Doing this carbon reduction, however, requires a lot of investment in zero carbon heating and cooling sources and improvements to building envelopes. However, where the energy sector must reach net-zero emissions by 2050, direct building carbon emissions must decrease by 50% and indirect building emissions from power generation by 60% by 2030. These efforts imply a decline in building sector emissions by approximately 6% per year from 2020 to 2030. For comparison, the global energy sector carbon emissions decreased by 7% during the Covid-19 pandemic.

When fulfilling targets to achieve the goals of the Paris Agreement, key actions – from high-efficiency lighting and zero carbon building design to low-cost building envelope measures – could result in considerable reductions in global energy savings and carbon emissions per year between 2020 and 2050. This requires a total change in the collaboration of all actors along the building value chain, including new policy signals to address the market, new business models, building product innovation, and innovative financing solutions. Figure 1.18 shows the roadmap by Global Roadmap for Buildings and Construction (Global ABC, 2020) which focuses on targets needed to achieve a zero carbon emission, efficient, and resilient building stock by 2050 in eight key areas: urban planning, new buildings, existing buildings, building operations, appliances and systems, materials, resilience, and clean energy. For each of these proposed key actions, targets for policies and technologies, and key enabling actions in the short-, medium- and long term are recommended to enable the delivery of these targets. More details about these key factors can be found in this study. The roadmap outlines a common vision for decarbonising the construction sector and buildings and supports the development of national or subnational strategies and policies. There are also regional roadmaps by the Global ABC, covering Africa, Asia, and Latin America.

To support the decarbonisation of new and existing buildings, effective policies and codes need to cover the entire life cycle, including the design, construction, operation, and deconstruction stages, and also act beyond site boundaries through neighbourhoods and energy supply. To promote such action, greater collaboration involving a range of stakeholders is needed, including the client, the client organisation, policy makers, urban planners, architects, consultants, construction companies, material suppliers, developers, and investors.

Different NGOs around the world are working with clients, developers, financing organisations, and building users to promote and accelerate net zero buildings. For example, the WGBC supports net zero buildings through its Advancing Net Zero (ANZ) project with a whole-life vision for total decarbonisation of the built environment. The ANZ framework includes guiding principles on how to reach net zero operational and embodied carbon. With the WGBC's schemes and programmes, it is possible to obtain an overall understanding of how to adapt suitable key concepts to the local market. At a regional level, GBCs are working on solutions to specific regional challenges, such as developing a foundational embodied carbon primer document for

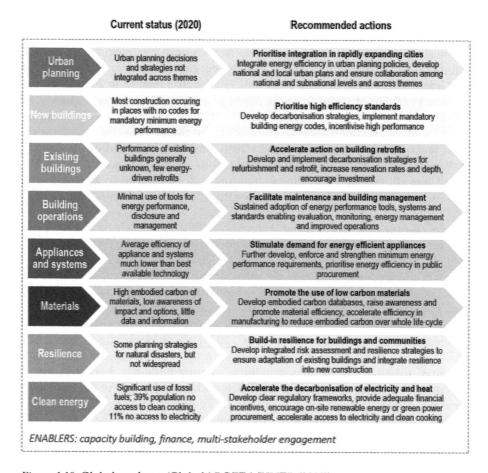

Figure 1.18 Global roadmap (GlobalABC/IEA/UNEP (2020)).

the Asia Pacific region, in Europe, implementing the BUILD UPON2 project developing national deep renovation strategies towards decarbonisation and energy efficiency, and the Cities Climate Action project in the Americas. To stimulate solutions for the market to deliver this vision and to send strong signals of industry demand to policymakers, the Net Zero Carbon Buildings Commitment recognises businesses, organisations, cities, states, and regions. They are taking action with an advanced path for net zero buildings by 2030, and/or enacting codes for new buildings to be net zero by 2030, and all buildings, including existing, to be net zero by 2050.

Next to commitments, organisations are also developing frameworks for net zero, such as the need for a holistic design approach to reach net zero. Such frameworks will ensure that energy needs are sufficiently reduced 'to permit renewable energy or zero carbon sources to meet most or all of the remaining space conditioning energy requirements'. Organisations contributing to the vision include the Building System Carbon Framework, C40 (Clean Construction Forum), Architecture 2030 (Achieving Zero), the Carbon Leadership Forum, and many more.

Regenerative development

Designers, engineers, and construction project managers mostly address the efficiency of the product, i.e. the buildings, through a green or sustainability concept, while failing to understand the systems they are trying to sustain. It is time to change this mental model to one that better reflects how our environment actually works and enables us to design and build with the whole system in mind. Whole systems and living systems thinking, which can help transform the way of practicing sustainability to linking the natural and built environments in a holistic manner, is crucial.

As early as the second half of the 2000s, discussions about regenerative development in the construction sector began. They largely continued during the first half of the 2010s, but when the focus shifted to the urgency of actions regarding climate change and the construction sector and the necessity of adapted LCA tools, the subject faded to less of a priority. From the formulation of UN SDGs and roadmaps to 2030, the subject is gaining more attention in the beginning of the 2020s. As Reed (2007) describes, 'to widen the view from only see the trees in the forest (level 1) to see the whole forest from outside (level 2) and, lastly, from above discover there are a lot of other forests around (level 3)'. Transforming this into a building concept – level 1 concerns efficiency – for example, energy efficiency or material productivity. Level 2 is about sustainability, effectiveness of the SDG's or the triple bottom-lines, and level 3 is about understanding the entire system, the purpose of sustainability, as shown in Figure 1.19. The figure presents the learning process from a design point of view – the pre-level of conventional practice with no breaking regulations or building codes – high-performance design is about a technical efficiency approach to design and may limit the benefits of the larger natural systems. Green design is considered a general term implying a direction of continual improvement towards an ideal of 'doing no harm' with a zero impact on the environment. This ideal is called sustainability.

Beyond sustainability begins an entire system mindset with a 'doing more good' approach. The first step is to use the activities of design and building to restore the capacity of local natural systems, a restorative stage of approach. The next step, reconciliation, involves humans in the design process as an integral part of the local nature, considering human and natural systems as one. Regenerative design is a design process that engages and focuses on the entire system of which we, the humans, are a part. The places, communities, watersheds, and bioregions are where humans can participate. By engaging all the key stakeholders and processes of the place, humans, other parts of biotic diversity, earth systems, and the consciousness that connects them, the design process builds the capability of humans and the 'more than human' participants to engage in a continuous relationship. The design process supports continuous learning through feedback, reflection, and dialogue so that all aspects of the system are an integral part of the process of life in that particular place to sustain sustainability.

Regenerative development is a place-based development of capabilities necessary for living systems to increase complexity, diversity, capacity to support all life and provide future options (i.e. health and well-being). Examples of frameworks for regenerative development are regenerative community development by the Regenesis Group's approach and the Living Environments in Natural, Social, and Economic Systems framework from the Centre for Living Environments and Regeneration.

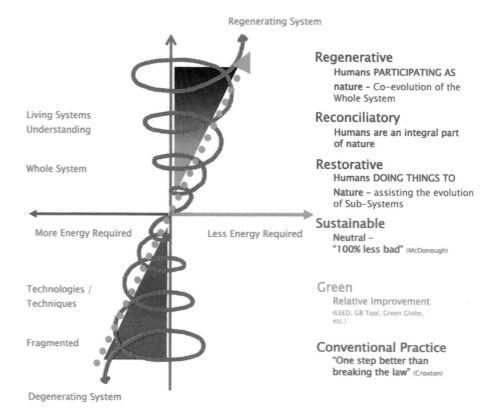

Figure 1.19 Regenerative development (Reed, B. (2007)).

Regenerative sustainability

Regenerative sustainability represents a necessary worldview and paradigm shift for sustainability. It includes and exceeds conventional sustainability by adopting a holistic worldview. Regenerative sustainability, as described above regarding regenerative development, sees humans and the rest of life as one system, with continual processes to consolidate the uniqueness of each place or community. The aim of regenerative sustainability is to confirm living systems in a fully integrated individual-to-global system. It calls for humans to adapt living systems principles of wholeness, change, and relationship, as nature does. The inhabitants of a place or community along with the stakeholders who developed it ultimately decide whether it is (un)sustainable or (not)thriving, but places are constantly changing. Developing communities that could continually confirm higher levels of health and well-being is necessary for regeneration. These abilities cover adaptation, self-organisation, and continual improvements as well as decisions about infrastructure, land use, governance, food systems, cultural practices, and lifestyles. Instead of seeing problems and solutions, regenerative sustainability sees living systems as existing in changing conditions of health and complexity. Regenerative sustainability could be more difficult to implement than conventional sustainability because of its more ambitious aims. Data from research and regenerative sustainability practitioners imply that regenerative approaches are more inspiring and motivational than conventional approaches and more effective at achieving these aims. Regenerative sustainability has been used in regenerative development, regenerative community development, and several regenerative design technologies. While the concept of regenerative sustainability has been articulated relatively recently, its principles have been applied in practices such as regenerative development, ecological and regenerative design, ecological planning, and regenerative agriculture. This offers insight into what is possible when a holistic view is adopted.

How to adopt regenerative sustainability is about to get more attention regarding the shift in the role of buildings and the changing construction process due to climate change. It is still in its infancy and faces several practical and operational concerns for clients and construction project teams, such as:

- Living systems thinking represents a too great leap to mainstream construction.
- The additional time needed in the early design stages to engage stakeholder input and client commitment is a major obstacle.
- The scale of construction projects at which most project teams work could be incompatible with the regenerative ambitions.

As of 2021, only a few regenerative development case studies have been published and most are in rural areas with small and quite affluent community groups. Urban and suburban areas are not homogenous in terms of diversity in social and economic equity, demographics, and closeness to service. Different neighbourhood communities within cities could also have qualitatively different engagements with place, relations, and capabilities. Regenerative development can be considered improper in more complex urban situations. Unless more case studies are conducted in these contexts, regenerative applications may be restricted to the outer edge of urban areas.

Regenerative community development

Regenerative community development is based on the existing approaches. Communities are defined as the biotic (the living factors that influence the environment) and abiotic (the non-living factors that influence the environment) components of the interaction of the complex webs of life. This contrasts with the neighbourhood, which indicates a geographically bounded area in a human-dominated system. Communities are the blocks of nature and societies, neighbourhoods, cities, landscapes, watersheds, and bioregions. Working with the structure of living systems at a community level could stimulate regeneration across all scales; for example, community action at the neighbourhood level stimulates change at the city level, which stimulates change at the landscape and bioregional level, and so on. Research and action at the community level is key to sustainability and regeneration and should be holistic. Its processes and tools help inhabitants better understand life-giving flows (e.g. water, food, energy, organisms, and information) through their community and the communities of which they are a part as well as their relationships. It transforms living systems principles and characteristics into general indicators and strategies for whole-system health and regeneration, specific to a place through ongoing processes. These processes help the inhabitants and stakeholders of a place to integrate ecological and sociocultural dimensions of living systems as well as development (e.g. participatory processes in urban design and landscape architecture) and products (e.g. ecological urban infrastructure, city plans and codes, buildings, and water systems) that are involved in community development. In short, regenerative community development seeks to develop regenerative communities that form the matrix, out of which all aspects of regenerative being, regenerative living, and regenerative whole living systems arise.

Of particular interest are the thousands of ecovillage communities around the world that have been developing ecologically and socially in the last decades. Although ecovillages most often fall into the category of regenerative design, many are attempting to shift the larger communities, which they are a part of, to create a bigger social change. They are living laboratories that provide years of data and invaluable insights into sustainable and regenerative living. The United Nations recognises ecovillages as the best strategy for achieving sustainability goals. Ecovillages could provide a springboard for increasing the efforts of community development to become regenerative on a larger scale.

Regenerative buildings

The differences between green and regenerative approaches are significant, where current green approaches already have many existing practical experiences and where regenerative approaches appear without sufficient track records to fully support the goals. As mentioned before, there is an urgent need to make significant global carbon emissions reductions during the decades before 2050, and the design stage of the construction process focusing on restricting and totally reducing carbon emissions will dominate future work, through proactive efforts, shifting of client demands and expectations, or more stringent building codes. However, as with past environmental issues, focus on climate change with the seriousness it deserves will likely be compromised by other, more locally pressing societal and political priorities. Furthermore, the construction process itself is not able to fundamentally shift to a regenerative approach. New strategies and contexts could probably be implemented partially and selectively within an already existing performance, depending on the involved parts' experience, capability, and commitment in zero or beyond zero carbon buildings. However, the necessary reductions in carbon emissions from the built environment have not yet been generated from green building strategies, and environmental performance has improved in recent decades.

As new infrastructure and buildings are made more resilient to a changing climate, adaptation to regenerative development will depend on people's day-to-day actions in the places they live. Unlike green building practices, regenerative approaches assign design professionals as co-learners and co-creators together with community members and other stakeholders, that is, those who will be most directly affected by climate change. However, many green building design strategies and technical knowledge remain valid when reducing other environmental impacts. While regenerative development gains more acceptance, green building knowledge and experience will need to be revised into more holistic thinking, a greater understanding of interactions between strategies and many qualitative factors.

Regenerative development and design are currently in the same position as the emergence of green buildings some decades ago but are unlikely to be mainstreamed in the same way or as fast. Non-governmental organisations' support, such as the Green Building Councils, must be revised to be on the same level as for green buildings. It will, of course, rely on the direct engagement and involvement of community members and stakeholders. Thus, they are probably more framed within an ecological worldview and adaptable to the whole system thinking, while the clients are not. Clients need to reconsider their task of managing and operating their building stock to a regenerative view and approach.

1.6 The big challenge of existing buildings

This textbook addresses issues concerning sustainability and regenerative development regarding existing buildings in situations where connecting the life cycle and circular view of the entire construction process is relevant. However, to begin, a basic overview of existing buildings and sustainability is provided.

In general, construction relates to upcoming developments or new buildings. However, to achieve sustainability, a sizable stock of existing buildings must be considered for sustainable and regenerative development. Globally, less than 5% of total building stock represents new buildings, whereas existing buildings comprise more than 95% (see Figure 1.20). Only a fraction of new buildings, which represent a fraction of a fraction of the total stock of buildings, are certified. Most of the available certification tools apply only to new buildings; however, some tools consider the entire or part of the construction process and include existing buildings. In addition, various upcoming tools consider the entire process.

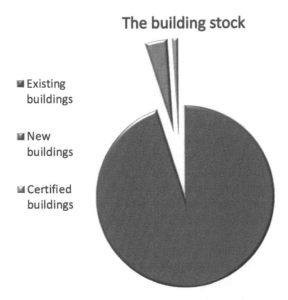

Figure 1.20 The proportions of buildings in the building stock (author original).

To reduce operational carbon emissions in existing buildings, refurbishment or renovation of the building envelope and its technical systems is crucial. Ultimately, they should reach the standards of new buildings. Operational carbon can also be reduced by ensuring higher quality during maintenance and refurbishment to extend the life of the building. In addition, reduction can be achieved through increased use of existing buildings and the following higher utilisation rates (especially by combining these measures).

Actions critical to achieving a sustainable building stock

According to the UNEP Global Status 2020, the following actions are critical for achieving sustainability in existing building stock.

- Integrate urban planning with the long-term goal of decarbonising buildings and the construction process. This should be done by implementing frameworks that include land use efficiency, green spaces, and district energy systems.
- Set distinct energy performance targets for existing buildings in collaboration with clients, users, and other stakeholders. Promote passive and bioclimatic design concepts to reduce energy demand. Develop national and local strategies to decrease the carbon emissions of existing building stock and increase the annual energy efficiency renovation rate to 4% by 2050.
- Enable dynamic rating systems of building performance that include energy performance with the possibility of comparison. Static rating systems only describe the performance at a specific time, as a 'snapshot'.
- Implement regular energy performance assessments to ensure that systems are being efficiently operated and maintained, e.g. usage of advanced smart energy management systems.
- Implement the use of the building renovation passport (see below for more information) as a system for regular information collection related to the building's technical system, operation, and energy use. The passport will support the availability and access of building information for current and future clients, users, owners, and future operation and maintenance organisations.
- Develop, review, and enforce minimum energy performance standards (MEPS) (see below) as a product quality and performance requirement. Expand and update the MEPS to encompass major appliances and systems and set the energy performance requirements for networked devices.
- Promote investment in high-performance systems for space and water heating, cooling, and ventilation and lighting, especially through procurement practices.
- Increase the demand for materials with low embodied carbon and promote the efficient use of low carbon energy in the production of major building materials.
- Adopt material efficiency strategies to promote circular economy concepts that use lifecycle approaches throughout the entire construction process to reuse construction materials and phase out the use of potential refrigerants that perpetuate global warming.
- Improve long-term resilience of building stock with increased use of risk assessments and resilience planning for emergency response. Consider future changes in climatic conditions related to flooding, wind, storm water and heat, and how buildings need to maintain resilience in the future. Map and build long-term strategies

for ways in which existing buildings must be adapted to mitigate extreme climate events.

- Promote the use of onsite and building-integrated renewable energy-producing devices, including solar PV, solar thermal, geothermal, micro-wind, and advanced biofuels where appropriate and possible. To support this, develop a distinct regulatory framework that defines operational rules, replacement schemes, and goals at the national and local levels.

Key principles regarding regenerative design and refurbishment of buildings include:

- Enabling positive interactions between people, technology, and nature to encourage post-occupancy behaviour
- Mitigating the effects of the surrounding microclimate and facilitating a comfortable interior environment with sufficient choice and use of new and existing materials
- Integrating natural systems in the building envelope to improve the health and well-being of its occupants and to restore local ecosystems
- Actively seeking alternative options for involvement in energy-sharing strategies and initiatives
- Redesigning building(s) to be adaptable for future changes of technology and social conditions

Building renovation passport

Over the past 20 years, energy efficiency standards and energy performance certificates (EPCs) have been introduced in different forms globally. The aim was to make the energy performance of individual buildings more transparent for use in a variety of contexts, including:

- Independent quality controls
- Declarations of eventual penalties for non-compliance
- Display of energy labels in advertisements
- Hand-outs for sale and rent transactions
- Improvements from renovation recommendations (cost-effective and cost-optimal measures)

However, ensuring efficiency and near zero carbon emissions in building stock by 2050 is a major challenge. The quality of the energy renovations of building stock is very important. Despite the proven economic and technical gains and the societal and environmental benefits of building renovation, rates of renovation remain very low and significantly below the expected level. Clients and investors continue to experience many barriers to improving the energy performance of their buildings. In addition to difficulties in financing, one of the most often quoted barriers is lack of knowledge regarding what to do, where to start, and the order in which measures should be undertaken. Even when some of the most important benefits of renovation – improved thermal comfort and air quality, more daylight entry, and improved health of occupants – are the main impetus for renovation, they are not covered by current EPC formats.

Most of the current EPCs provide a snapshot of a building's performance at a given time and lack consistent recommendations regarding the necessary steps to bring the building to a zero energy building standard in the future. Clients require easily available and reliable information to drive their investment decisions and, regarding renovation, more complementary information with the EPCs to provide a long-term, step-by-step renovation. For decades, the idea of introducing a building passport (BP) with the aim of improving the quality of data communication between the different stakeholders involved in the renovation process has been one of the main issues discussed; the objective is to provide information to the client, potential buyers, investors, customers, or users of the building. Currently, there is no common definition of BP. While suggestions for a renovation roadmap for a specific building are based on quality criteria, a building renovation passport (BRP) is defined as a document that includes a long-term (up to 15 or 20 years) step-by-step renovation roadmap for a specific building. Based on an onsite energy assessment concerning specific quality criteria and/or indicators established during the design phase of the renovation process and through dialogue with the client, it outlines relevant measures and renovations that can improve energy performance. The expected benefits of reduced heating bills, comfort improvement, and carbon emission reduction are a basic part of the BRP and should be explained to users in a user-friendly manner. This renovation roadmap can be combined with a logbook of building-related information on aspects such as energy consumption, executed maintenance including material exchange and, if suitable, how energy production relates to the actual building.

The first step towards a BRP begins with an onsite data assessment of the actual building, performed by external auditors, the client, and the tenants. Based on the actual EPC, it includes documentation of executed refurbished work, age, and characteristics of installed equipment, including technical conditions and suitable monitoring systems of the building. The outcome of the assessment is a comprehensive step-by-step renovation roadmap with tailored solutions to achieve deep-staged renovation. The roadmap should comprise a renovation plan with a horizon of 15–20 years that considers the entire building, suggesting selected measures in a certain order to prevent the need for any additional measures during the execution.

There is no common definition for deep renovation, staged renovation, or deep-staged renovation. However, all initiatives share common features, such as the aim to increase the level of achieved energy performance, to ensure consistency between short- and long-term measures and to align the target for the performance of individual buildings with the long-term target for the entire building stock.

By supporting staged renovations adapted to client demands, BRPs offer the opportunity to maintain an overview of the full range of renovation options and to easily identify each renovation step from beginning to end. Consequently, staged renovation strategies promote the client's investment through a deeper renovation process, especially when specific elements that must be considered during future renovations are also emphasised (e.g. roof insulation, roof overhangs, downspout connections, adjustment of the boiler and piping penetrations for future solar systems). The final product comprises a renovation roadmap that describes each renovation step and the links between implemented measures, presenting the renovation as an improvement plan for the entire building that includes indoor quality and helps to avoid lock-in effects. It is essential for clients to take control of the project – for example, for clients with a single building, the uncertainty of future renovation options typically leads to hesitation regarding renovation decisions or to limiting the renovation project with easier measures. From this perspective, any tool that incites a long-term perspective and allows clients and investors to obtain a clear overview of long-term renovation plans with short-term, adaptive, and flexible measures (e.g. sequencing of measures' installation over time) could improve the value of the renovation and increase the client's confidence in the decision. BRPs could provide a comprehensive set of relevant indicators (e.g. energy consumption, carbon emissions, thermal and acoustic comfort, indoor air quality, and daylight), including a dynamic way to disseminate information about recommended detailed improvement strategies over time to stimulate deep or deep-staged renovations.

In addition, the BRP can include a logbook, where information about building features (e.g. stability, durability, water, installations, humidity, maintenance requirement, etc.) can be consistently collected and updated to establish a recorded base of information and data related to the specific building. The logbook could also include other aspects of information related to each individual building, such as financing options available for renovation projects (e.g. green loans, incentives, tax credits), energy bills, equipment maintenance recommendations, and insurance and property obligations. The client is the primary user of the logbook, and depending on its intended use, can grant access to select information available to public authorities (e.g. municipality, property tax office), consultants, contractors, and users, while keeping other data restricted or semi-public upon authorisation to third parties. The logbook could also be used as an interactive tool to monitor – both at the individual building level and building stock level – and compare real energy consumption with supposed energy consumption, sending alerts in cases of unusual consumption patterns or flaws in technical installations. Further, it could provide information regarding the requirements of certified contractors or installers, facilitate invoicing, and simplify the process for subsidies or loan repayment.

Minimum energy performance standards

Minimum energy performance standards (MEPS) are regulated by minimum standards for energy use in, or carbon emissions from, existing buildings. MEPS require the improvement of buildings to meet a specified standard at a chosen trigger point or date and can include standards that become more restrictive over time. As such, MEPS drive the desired target and the required depth of renovation. The example in Figure 1.21 illustrates MEPS that require buildings to achieve a certain performance, expressed as energy performance consumption per square meter. The level of required performance increases over time, with the expectation that the target for the building stock will be in the best class by 2050. MEPS include regulations that require buildings to meet a minimum performance standard, in terms of a carbon or energy rating or minimum renovation measures, by a specified deadline or at a specified moment in the life cycle of the building. MEPS can be applied to the entire building stock or to specific sectors, tenures, building types or sizes, or different types of clients. Even though MEPS establish regulated standards for buildings, building designs depend on local and national codes, building stocks and targets. The regulations define a standard to be achieved (what), a section of the stock to be targeted (where), and the schedule for performance (when). Different variants of the MEPS concept are already in use worldwide, and practical outcomes show improved results across the targeted stock, high levels of compliance when the MEPS framework is in place, and a goodwill signal effect on the markets.

The most common units used include carbon, measured in CO_2 per surface unit (e.g. square meter), or energy ratings, such as kilowatt-hour per surface unit (e.g. square meter) or an EPC class. The ratings used can be either asset-based or operational. A small number of minimum requirements are defined as the presence of a minimum set of building HVAC system measures. Regulations could also be established to eliminate the worst-performing buildings, for example, by setting a minimum EPC class, or by targeting a larger proportion of the building (e.g. buildings with below-average performance). The reviewed MEPS could address different building stock sectors, tenures, and different types of clients or be based on building type or size. The trigger for implementation could be a hard date by which all obligated buildings must comply, accompanied by a time frame for gradual improvements. Implementation could also be based on a date that coincides with one of the trigger points in a building's life cycle, such as major renovation, inspection, or auditing, different transactions such as sale, change of rental contract, or new tenants. Other useful trigger points could include repair, planned maintenance, building extension, kitchen and bathroom replacements, and other general work. Although the MEPS specifically target energy or carbon performance in most cases, other countries are seeking improvement of the building stock from a more holistic perspective, e.g. Australian and New Zealand

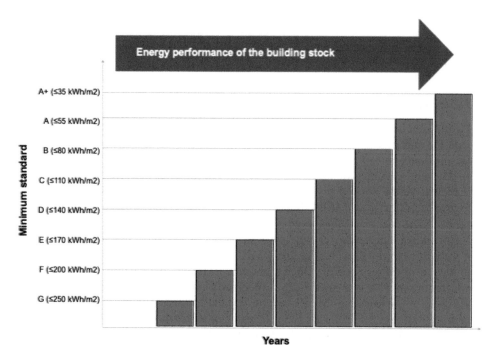

Figure 1.21 MEPS required performance over time (Sunderland, L. and Santini, M (2020)).

standards and the UK standard for social housing are based on a more expansive set of health-based requirements, including minimum energy standards. Some MEPS target parts of the building stock to deliver on local priorities, e.g. various MEPS in the United States aim to deliver on city-based carbon targets, focused on achieving the greatest carbon savings by targeting large commercial buildings. MEPS in the United Kingdom and New Zealand include requirements that target housing, specifically private, rented housing with a focus on improving housing standards, energy affordability, and health and well-being. MEPS could be integrated with a roadmap for decarbonisation of the building stock and with a more expansive decarbonisation strategy.

Roadmap to a decarbonised building stock

Many countries have established a comprehensive plan that provides a pathway to a decarbonised building stock by 2050 by way of national long-term renovation strategies. Long-term renovation strategies provide an overview of the building stock and establish a roadmap with measurable indicators of progress. This strategy could include milestones for 2030, 2040, and 2050. One example is the previously mentioned BRP. However, MEPS can also create a bridge between national building sector decarbonisation targets and an individual building achieving those targets, e.g. long-term goals and gradually higher performance standards established in the United Kingdom and Scotland. Others have introduced MEPS at a minimum level and committed to a review of the standards, e.g. on a five-year basis, with the possibility of increasing the requirement steps. The first approach is likely stronger because it provides a clearer route to the main decarbonisation target. Many MEPS are based on energy units or carbon units, such as in the European Union, where EPCs are based on a single energy rating or a combination of energy, cost, and carbon ratings. With the standard based on a minimum carbon rating, the focus falls on the decarbonisation of the used energy. When building energy performance is the main issue, more focus is placed on the building's envelope and installation systems than on the carbon content of energy. This has implications for the MEPS design. When based on an energy unit, it can be linked with measures of the carbon intensity of the energy delivered. If suitable, energy units can also be linked or replaced with carbon units, as building rating systems are developed or improved over time.

In addition to different variations of MEPS and a broader concept regarding decarbonisation, the efficiency of heating and cooling systems of buildings can be improved. For example, use of buildings with new or expanded district heating and cooling systems can be considered. Clients must be able to plan when complying with MEPS that encompass low-carbon or decarbonised district systems.

Another consideration involves improving the electrification of buildings to a more carbon-reducing system. Measures could be taken to enable individual buildings to electrify efficiently on-site through photovoltaic or wind power or by connecting to and interacting with a grid with a renewable mix. Fuel switching, building-level heat and power storage, grid switching, and demand response measures may not be driven solely by MEPS.

In practice, some MEPS developers are exploring a variety of strategies and supportive tools that consider deeper decarbonisation and sustainability objectives. For example, Scotland is weighing the introduction of MEPS that consider both energy and environmental (carbon) units. The Brussels-Capital Region in Belgium is developing BRPs and improved EPCs to provide recommendations on measures with more content than ordinary energy efficiency and efficient heating systems, including several sustainability measures.

Case 1: Lessons learned from a regenerative renovation, Spain

This case study focuses on an existing residential building that has undergone a process of regenerative refurbishment, with the goal of becoming the first regenerative building in Spain. Located in a small village on the slopes of the mid-Pyrenees approximately 220 km northwest of Barcelona, the building experiences an average local annual temperature of 12.3°C and approximately 700 mm of rainfall per year. Situated on top of a hill overlooking a nearby river and surrounding mountains, the building is exposed to the sun year-round and to a continuous upward wind from the west.

The site was chosen due to its orientation regarding sun exposure during the entire year (for passive solar heating) and the beneficial wind direction, which enables natural ventilation and passive cooling. The construction process for the refurbishment project started with a pre-design phase, followed by a design phase, both of which lasted for six months. Construction began in the spring of 2017; major work was completed in the spring of 2019, and it began to operate in the autumn of 2019. To obtain regenerative status, the building was registered to a regenerative certification organisation (see Chapter 7) and its operation and performance had to be monitored over a 12-months period to verify water and energy consumption. The actual certification should be completed in 2021. Thus far, it seems to be net positive in terms of both energy and water consumption.

The two-floor building is oriented east-west, attached to another property at its east wall, and has its longer façade exposed to the south. The ground floor space is divided into a living room, dining room, and kitchen with an orientation to the south, while the stairs, hallway, and bathrooms are situated on the north side. The upper floor comprises four bedrooms oriented to the south and two bathrooms, one near the east and one near the west side.

A solar over-heating protection strategy has been implemented with an overhang that blocks the summer sun and utilises the beneficial winter sun for passive solar heating. In addition, deciduous plants and trees have been planted adjacent to the south and west sides to prevent over-heating. This integration of plants deepens the connection of the building with its surroundings, making it a part of the local bio-system. Additional plants on the south side include kiwi and grape vines, planted near the building and used for local food production. Although they do not provide direct shade for the building, they create a shade canopy for the users of the southern courtyard during the summer period. The use of these plants is an example of multiple ecosystem services (see Chapter 4). Kiwi and grape vines provide shade in the summer and allow adequate sun to get through for passive solar gain in winter, provide food, add a biophilic aspect to the building, and offer falling leaves as compost to improve the soil.

The building is insulated by wood fibres on the interior of the outer walls, roof, and floor, thus lowering the thermal mass. Furthermore, the inner layers of the walls include straw and clay. The building is protected from overheating by using the solar protection strategy mentioned above. Natural ventilation is sufficient for comfortable indoor conditions during summer. Controlled mechanical ventilation and high thermal insulation provide internal comfort during the winter. The HVAC system purifies stored rainwater for drinking water and greywater use. The devices used include composting toilets, low-flow fixtures, efficient appliances, and a special low-flow shower system. Urine and faeces are separated and stored for later use as fertilisers. Greywater from the kitchen, bathrooms, and sinks is recycled and purified through a constructed natural wetland for later irrigation use. Water consumption is approximately 63 litres per year and person compared to the average use of 150 litres. The installation of solar PV panels on the roof of the building and battery storage for peak use resulted in a positive annual energy balance with a deficit during the autumn and a surplus during the rest of the year. The use of local materials on the exterior of the building, as well as in the interior, is a challenging task due to the availability, type of material, and content of hazardous components. A good example of biophilic design (see Chapter 6) is the integration of a 'river' that runs through the kitchen consisting of a small floor channel that directs the rainwater from the roof towards the rainwater capture tanks in the basement.

Various conclusions were drawn from this case. When working with regenerative building and renovation, the design needs to include certain principles to ensure that the outcome is as cost-efficient as possible. One of these design principles is the stacking of functions. The use of biophilia (see Chapter 6) as a driver for design is another important principle that needs to be taken into consideration. One of the concerns related to more frequent regenerative building projects is higher investment costs, increased by additional plumbing systems, PV panels, and especially energy storage systems. To stimulate implementation of the regenerative concept and reduce the increase in investment costs, an incentive – such as local or general construction codes – must be established by authorities. The outcome could be to reduce certain design aspects, materials, or technologies. Moreover, energy and water policies can influence use of the regenerative concept. When implementing passive solar design principles and fulfilling the requirements of Living Building Challenge certification (the 'imperatives'; see Chapter 7), structured project management must be established and followed. It is also crucial to establish the project team from the starting phase of the project until its completion, i.e. from the pre-design and design phase to the construction and use phases. Project management must begin prior to the planning phase and include a thorough assessment of the context and a visioning phase to determine the project's objectives. Moreover, a project advisor or mentor who will support both the team and individuals must be assigned. The establishment of an integrated project team with clearly defined project objectives enables the delivery of appropriate solutions which, otherwise, could not be achieved due to a limited perspective, lack of knowledge, or even lack of will.

Ordinary design-bid-build contracting is not suitable for a project that demands numerous specialists, engineers, and consultants with different profiles. Instead, a design-build-contract and project delivery process is more suitable for the efficient implementation of regenerative goals (see Chapter 6). Furthermore, cost analysis and comparison with a traditional construction process would be valuable for further evaluation; such an analysis would elucidate not only the construction costs but also the contribution of the project to the natural, human, social, and constructed capital.

Implementation of the regenerative certification standard in urban settings can be challenging due to existing urban planning regulations and conditions. However, because of the criteria demand, the integration of the regenerative certification design principles can substantially improve the quality of urban life. By fulfilling the 'Place petal' (see Chapter 7 concerning the regenerative certification Building Living Challenge), a substantial contribution can be made towards improving and preserving biodiversity in the cities' neighbourhoods as well as stimulating urban agriculture and human-powered living as strategies to mitigate climate change and reduce urban heat island effects and sustainable growth. Projects that operate within the water balance of a given place and climate can facilitate reconsidering and redefining how people use water and wastewater, and how rainwater is harvested. These aspects are important for improving the resilience of habitats, considering climate change and the possibility of extreme droughts or violent storms. At times, the fulfilment of criteria regarding energy, PV, and materials is challenged by local regulations. However, with advancements of these technologies and different refurbishment strategies, they can be more easily achieved in urban settings in the future. Considering that the regenerative certification standard is largely composed of qualitative criteria, the holistic integration of the project team with the certification and construction teams, and collaboration with the occupants, craftsmen, and even local municipalities ensures adequate and collective fulfilment of the qualitative criteria, thus delivering a regenerative building in its entirety. The knowledge gained from the unique experience of regenerative design and the construction of the presented case study can be applied in future regenerative buildings to increase their wider acceptance and implementation in contemporary construction practice.

Bibliography

Berchtold, M. (2020) *Meeting the dual challenges of Covid-19 and climate change*, practitioners session D1 O2 S1, presentation, Gothenburg, Sweden, November 20 Beyond 2020–2002.

Bergman, Z. et al. (2018) *The Contribution of UNESCO Chairs toward Achieving the UN Sustainable Development Goals*, Sustainability 2018, 10, 4471.

Boers, N. (2021) *Observation-Based Early-Warning Signals for a Collapse of the Atlantic Meridional Overturning Circulation*, Nature Climate Change 2021, 11, 680–688.

Bon, R. and Hutchinson, K. (2000) *Sustainable Construction: Some Economic Challenges*, Building Research and Information 2000, 28(5/6), 310–314.

Center for Sustainable Systems (2020) *Wind Energy Factsheet*, University of Michigan. Pub. No. CSS07-09.

Cole, R.J. (2020) *Navigating Climate Change: Rethinking the Role of Buildings*, Sustainability 2020, 12, 9527.

Constantino, N. (2006) *The Contribution of Ranko Bon to the Debate on Sustainable Construction*, Construction Management and Economics 2006, 24(7), 705–709.

Craft, W. (2017) *Development of regenerative design principles for building retrofits*, Proceedings World sustainable built environment conference, WSBE17, Hong Kong, 2017.

Crucifix, M. (2016) *Earth's Narrow Escape from a Big Freeze*, Nature 2016, 529, 162–163.

du Plessis, C. (2007) *A Strategic Framework for Sustainable Construction in Developing Countries*, Construction Management and Economics 2007, 25(1), 67–76.

Eurostat (2020) Waste statistic 2018, https://ec.europa.eu/eurostat/statistics-explained/index. php?title=Waste_statistics&oldid=503336#Total_waste_generation, access 2021-04-06.

Fabbri, M. et al. (2016) *Building Renovation Passports – Customized Roadmaps towards Deep Renovation and Better Homes*, Building performance institute Europe (BPIE), Brussels, Belgium, 2016.

Falk, J. et al. (2020) *Exponential Roadmap 1.5.1*, Future Earth, Sweden, January 2020.

Ganopolski, A. et al. (2016) *Critical Insolation-CO_2 Relation for Diagnosing Past and Future Glacial Inception*, Nature 2016, 529(7585), 200–208.

Gibbons, L.V. (2020) *Regenerative—The New Sustainable?* Sustainability 2020, 12, 5483.

GlobalABC/IEA/UNEP (Global Alliance for Buildings and Construction, International Energy Agency, and the United Nations Environment Programme) (2020) *GlobalABC Roadmap for Buildings and Construction: Towards a Zero-Emission, Efficient and Resilient Buildings and Construction Sector*, IEA, Paris, 2020.

IPCC (2019) Summary for Policymakers. In: IPCC Special Report on the Ocean and Cryosphere in a Changing Climate [H.-O. Pörtner, D.C. Roberts, V. Masson-Delmotte, P. Zhai, M. Tignor, E. Poloczanska, K. Mintenbeck, A. Alegría, M. Nicolai, A. Okem, J. Petzold, B. Rama and N.M. Weyer (eds.)]. In press.

IPCC (2021) Summary for Policymakers. In: Climate Change 2021: The Physical Science Basis. Contribution of Working Group I to the Sixth Assessment Report of the Intergovernmental Panel on Climate Change [V. Masson-Delmotte, P. Zhai, A. Pirani, S. L. Connors, C. Péan, S. Berger, N. Caud, Y. Chen, L. Goldfarb, M. I. Gomis, M. Huang, K. Leitzell, E. Lonnoy, J.B.R. Matthews, T. K. Maycock, T. Waterfield, O. Yelekçi, R. Yu and B. Zhou (eds.)]. Cambridge University Press, Cambridge, United Kingdom and New York, NY, USA, pp. 3–32.

Jonasson, J. et al. (2020) *Knowledge management of sustainable construction processes proceedings*, World Sustainable Building Conference, SB20, Gothenburg, Sweden, Beyond 2020.

Meyers, D. (2005) *A Review of Construction Companies' Attitude to Sustainability*, Construction Management and Economics 2005, 23(8), 781–785.

Pawlowski, A. (2008) *How Many Dimensions Does Sustainable Development Have?* Sustainable Development 2008, 16, 81–90.

Pearce, D. (2003) *The Social and Economic Value of Construction, the Construction Industry's Contribution to Sustainable Development 2003*, The construction Industry Research and Strategy Panel, nCRISP, Davis Langdon Consultancy, London, UK, 2003.

Persson, U. (2009) *Management of sustainability in construction works*, PhD thesis, Division of Construction Management, Lund University, Lund, Sweden, 2009.

Persson, U. (2011) *Procurement and sustainability in construction works – Two cases of construction works projects with sustainability demands regarding the phase of procurement*, Proceedings, World Sustainable Building Conference, SB11, Helsinki, Finland, 2011.

Persson, U. et al. (2008a) *Sustainable construction management at a project level – A modified environmental management system structure*, Proceedings, World Sustainable Building Conference, SB08, Melbourne, Australia, 2008.

Persson, U. et al. (2008b) *Ten years of sustainable construction – Perspectives from a North construction manager and a South architect point of view*, Proceedings, World Sustainable Building Conference, SB08, Melbourne, 2008.

Persson, U. and Olander, S. (2004) *Methods to estimate stakeholder views of sustainability for construction projects*, Proceedings, Plea 2004, The 21th Conference on Passive and Low Energy Architecture, Eindhoven, The Netherlands.

Petrovski, A. et al. (2021) *Implementing Regenerative Design Principles: A Refurbishment Case Study of the First Regenerative Building in Spain*, Sustainability 2021, 13, 2411.

Reed, B. (2007) *Forum: Shifting from 'Sustainability' to Regeneration*, Building Research & Information 2007, 35(6), 674–680.

Rockström, J. et al. (2017) *A Roadmap for Rapid Decarbonization*, Science 2017, 355(6331), 1269–1271.

Sesana, M. et al. (2020) *Overview of the Available Knowledge for the Data Model Definition of a Building Renovation Passport for Non-residential Buildings: The ALDREN Project Experience*, Sustainability 2020, 12, 642.

Steffen, W. et al. (2018) *Trajectories of the Earth System in the Anthropocene*, Proceedings of the National Academy of Sciences of the United States of America 2018, 115(33), 8252–8259.

Sunderland, L. and Santini, M. (2020) *Filling the Policy Gap: Minimum Energy Performance Standards for European Buildings*, Regulatory Assistance Project (RAP), Brussels, Belgium, June 2020.

United Nations, Department of Economic and Social Affairs, Population Division (2019) World Population Prospects 2019: Highlights (ST/ESA/SER.A/423).

United Nations Environment Programme, UNEP (2020) *2020 Global Status Report for Buildings and Construction: Towards a Zero-emission*, Efficient and Resilient Buildings and Construction Sector, Nairobi, Kenya, 2020.

United Nations, General Assembly (1992) Rio Declaration on Environment and Development, A/Conf. 151/26 (Vol 1) Report of the United Nations Conference on Environment and Development, Annex 1.

Vollset, E.S. et al. (2020) *Fertility, Mortality, Migration, and Population Scenarios for 195 Countries and Territories from 2017 to 2100: A Forecasting Analysis for the Global Burden of Disease Study*, The Lancet 2020, 396(10258), 1285–1306.

WMO Greenhouse Gas Bulletin (2020) Can we see the impact of COVID-19 confinement measures on CO_2 levels in the atmosphere, No 16, 23 November 2020.

WMO Greenhouse Gas Bulletin (2021) The state of greenhouse gases in the atmosphere based on global observations through 2020, No 17, 25 October 2021.

WorldGBC (2021) Green building: Improving the lives of billions by helping to achieve the UN Sustainable Development Goals, https://www.worldgbc.org/news-media/green-building-improving-lives-billions-helping-achieve-un-sustainable-development-goals, access 2021-04-06.

2 The end? – circularity and deconstruction

2.1 Circular economy

Current trends in economic thought tend to be linear, that is, to take material and resources from the Earth's crust, convert them into products, use the product until it is worn out, and then dispose of it as waste. However, by maintaining the value of resources and materials, keeping them in circulation, thereby reducing material extraction from the Earth's crust, the linear mode of economic thought can transform into one that is circular.

The principle of circularity can be divided into four basic theses, similar to the laws of thermodynamics.

1 Matter and energy cannot disappear or be renewed. They can be transferred to other states of matter or energy, but with no energy loss.
2 While matter and energy tend to expand, their entropy stays the same.
3 The value of matter is significant for its concentration and structure. The value, or quality, increases with an increase in concentration, purity, and structure.
4 If the value of matter must be increased or maintained on Earth, energy must be obtained from beyond Earth, i.e. directly from the sun via radiation, or indirectly by photosynthesis, or geothermic or gravitational means, such as wind and tide.

According to the principle of circularity, the plan and design of society implies a decreasing use of material from finite sources and not exceeding their natural growth cycle when using material from renewable sources. The process of circular economy includes a limited ability to break down unwanted recourses, i.e. the resilience of Earth, and managing all resources and materials that circulate naturally from a long-term perspective, so that biodiversity is not depleted.

Although circular economy has different definitions, most support sustainable development and a regenerative or restorative approach. The Ellen MacArthur Foundation (2022) defines a circular economy as follows:

> a systems solution framework that tackles global challenges like climate change, biodiversity loss, waste, and pollution. It is based on three principles, driven by design: eliminate waste and pollution, circulate products and materials (at their highest value), and regenerate nature.

This is supported by the transition to renewable energy sources and materials. Transitioning to a circular economy requires disconnecting economic activity from

DOI: 10.1201/9781003177708-2

consumption of finite resources. This implies a shift to long-term resilience, potential business and economic opportunities, and environmental and societal benefits.

The first principle mentioned above – to eliminate waste and pollution – implies that waste is treated as a flaw of design; that is, that any design specification in a circular economy should ensure that materials re-enter at the end of use to be used again in a new way. The linear flow of material transitions to a circular flow, as shown in Figure 2.1 (right blue wing of the 'butterfly diagram'). Many products can be circulated by being maintained, shared, reused, repaired, refurbished, remanufactured, and, as a last resort, recycled. Food and other biological materials that are safe and harmless can regenerate nature and land, fuelling the production of new food and materials (see Figure 2.1, left green wing). With this focus, the concept of waste can be eliminated. Although it, sometimes, seems that waste is inevitable, waste is the result of design choices. Nature does not leave any waste; it is a concept introduced by humans and has been designed without asking: What happens to this at the end of its life? By adopting this first principle of circular economy, the material loop can begin to close and decrease the immense amount of waste that goes into landfills and incinerators every day. By focusing on upstream design, waste can be stopped before it is even created.

The second principle of circular economy – to circulate products and materials (at their highest value) – implies keeping materials in use as manufactured products or, when this is not achievable, as components or raw materials. Nothing becomes waste, and the inherent value is retained in the materials or products. The two fundamental cycles for maintaining materials and products in circulation include the technical cycle and biological cycle. In the technical cycle, products are reused, repaired, remanufactured, and recycled. In the biological cycle, biodegradable materials are returned to the earth through composting or anaerobic digestion processes. When a product can be circulated successfully, either via a technical or biological cycle, it is essential to design the product to perform with its circulation possibilities in mind. Many products exist in the current linear economy that cannot be circulated in either cycle or, consequently, end up as waste.

As presented in the right blue wing in Figure 2.1, the concept of the technical cycle is as follows:

- The highest product efficiency of retaining value is to maintain and reuse the products in their entirety to maximise the value. One way of prolonging the value is to adopt a sharing economy: different users who do not own the product get access to it, enabling it to be used by more people over time, and consequently, the value is retained. Moreover, prolonging the extent of reuse could involve resale and different cycles of maintenance, repair, and refurbishment.
- When a product can no longer be used, its components can be re-manufactured. Parts that cannot be re-manufactured can be broken down into constituent materials and recycled. Recycling is the last option because it means that the embedded value is lost, which is vitally important as the final step that allows materials to remain in the circular economy.

As presented in the left green wing in Figure 2.1, the concept of the biological cycle is as follows:

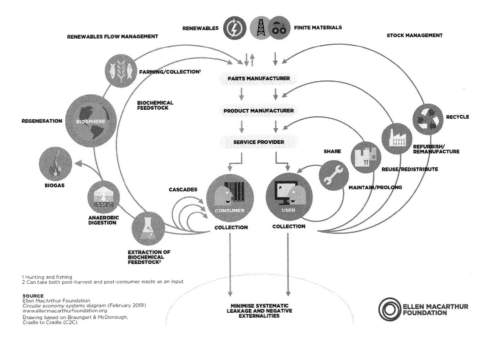

RENEWABLES   FINITE MATERIALS

RENEWABLES FLOW MANAGEMENT

STOCK MANAGEMENT

FARMING/COLLECTION¹

PARTS MANUFACTURER

BIOCHEMICAL
FEEDSTOCK

PRODUCT MANUFACTURER

RECYCLE

REGENERATION   BIOSPHERE

SERVICE PROVIDER

REFURBISH/
REMANUFACTURE

SHARE

REUSE/REDISTRIBUTE

BIOGAS

CASCADES

MAINTAIN/PROLONG

ANAEROBIC
DIGESTION

CONSUMER   USER

COLLECTION   COLLECTION

EXTRACTION OF
BIOCHEMICAL
FEEDSTOCK²

1 Hunting and fishing
2 Can take both post-harvest and post-consumer waste as an input

SOURCE
Ellen MacArthur Foundation
Circular economy systems diagram (February 2019)
www.ellenmacarthurfoundation.org
Drawing based on Braungart & McDonough,
Cradle to Cradle (C2C)

MINIMISE SYSTEMATIC
LEAKAGE AND NEGATIVE
EXTERNALITIES

ELLEN MACARTHUR
FOUNDATION

Figure 2.1 Circular economy systems diagram as the 'butterfly diagram' (Ellen MacArthur
Foundation (2019)).

- Biodegradable materials that cannot be reused, such as food by-products, can be
  circulated back into the economy in the biological cycle by composting or anaerobic digestion of organic materials.
- Valuable nutrients such as nitrogen, phosphorous, potassium, and micronutrients
  can be used to regenerate land to grow more food or renewable materials such as
  cotton and wood.
- Some products, such as cotton clothing and wooden furniture, can be circulated
  through both technical and biological cycles. Through the technical loop, it is
  possible to maintain, reuse, repair, and, sometimes, even recycle. However, it is
  also an option to return products to the biological cycle, to be composted, or
  anaerobically digested, to grow new cotton or wood.

The third principle – regenerating nature – implies a shift in focus from extraction
to regeneration. Instead of continuously degrading nature, practices are applied to
reinforce natural capital and allow nature to rebuild soils, increase biodiversity, and
return biological materials to the earth. Currently, most of these materials are lost
after use, and the land used for growth is depleted of nutrients. Waste is a human
invention. When a leaf falls from a tree, it is reintegrated into the forest. Natural systems have regenerated for billions of years.

The food industry is the obvious place to start when shifting to a circular economy
that regenerates nature. Food production is a significant driver of both climate change
and biodiversity loss. It depends on a large quantity of synthetic fertilisers, pesticides,

fossil fuels, freshwater, and other finite resources. These are the sources of pollution and damage to ecosystems and human health.

Regeneratively producing food focuses on improving the soil health. Regenerative farming practices can significantly reduce greenhouse gas emissions from food production by reducing dependence on synthetic inputs and by building healthy soils that absorb rather than release carbon. In addition to helping restore the natural carbon cycle, healthy soils have a better ability to hold water, reducing the impact of droughts, and a better ability to absorb water, thereby reducing the risk of flooding.

Regenerative food production practices include agroecology, conservation agriculture, and agroforestry (growing trees around or among crops or pastures). This results in agricultural land that more closely resembles natural ecosystems, such as forests and native grasslands, providing habitats for a wide range of organisms and increasing biodiversity. By reducing the need for synthetic inputs and pesticides, pollinators and microbes in soils, which are essential for the maintenance of healthy ecosystems, can thrive. Additionally, there are other benefits to natural ecosystems from the adoption of a circular economy. By keeping products and materials in use, less land is required to source virgin raw materials, such as mines. If we gradually disconnect economic activity from material extraction by keeping materials in circulation after use, an increasing amount of land can be returned to nature, and rewilding can occur. In a circular economy, land dedicated to material sourcing will increasingly focus on renewable resources, grown in a regenerative way, rather than the extraction of finite materials, which will continue to remain in circulation. This will be underpinned by a transition to renewable energy produced using infrastructure designed for reuse, repair, remanufacturing, and recycling. Transitioning to renewable energy alone will only minimise 55% of global greenhouse gas emissions. The remainder originates from how products and food are made and used and how land is managed. This is where the circular economy comes in.

In addition to these three principles, a framework composed of six levers was defined. Represented by verbs, the first letters of the six levers combine to create the acronym ReSOLVE. They outline the six actions needed to transition from a linear to a circular economy. The content of ReSOLVE includes the following:

- Regenerate
- Share
- Optimise
- Loop
- Virtualise
- Exchange

A more detailed explanation of these actions in the context of the built environment is provided in the next section.

2.2 Circular economy in the built environment

Typically, a building is constructed for a supposed economic life of approximately 50 years or more. However, the variety of usage determines real longevity. The estimated lifespan of a residential building is between 70 and 100+ years, whereas the lifespan of a logistics or industrial building could be estimated to last 30–40 years.

Such a difference is not related to the structural durability of the building but rather to the economic profitability of buildings in terms of satisfying the needs of the client. Therefore, logistics or industrial buildings are often deconstructed or demolished long before they reach their supposed construction lifespan. Generally, buildings are often inflexible and unadaptable to changes for future and unknown purposes; even buildings that are in good condition often undergo heavy renovations or are deconstructed for new construction. Therefore, common and general deconstruction practices only strengthen the linear economy and increase environmental impact. Consequently, deconstruction can still be a sensitive issue because common practices can take time to alter when clients or project stakeholders are not aware and do not enact strategies and communicate regarding an alternative practice.

In the context of the circular economy described previously, buildings can be preserved through regular maintenance, restoration, and renovation activities instead of being deconstructed. Deconstruction of a building is the last preferable solution if the building not offers adequate structural and sanitary conditions. Moreover, when the stage of deconstruction in the building's life cycle is imminent, a selective deconstruction process should be established for dismantling and disassembling because a product or material without damage can be reintegrated into the cycle of use; resources can be better separated; and the reuse, recovery, and recycling loops can be optimised.

Identifying reuse options and secondary materials to be recovered or recycled from a deconstruction project and to prevent waste requires direct action at the source of the deconstruction sites. A selective deconstruction process of dismantling and disassembling building products and materials must be implemented. This process comprises a sequence of activities for separating and sorting building products and materials. The producer or holder of construction and deconstruction waste coordinates the sorting of waste onsite. When waste is not treated onsite, it should be collected separately – especially wood, mineral fractions, metal, glass, plastic, and plaster – for recycling purposes. A successful selective deconstruction process requires rigorous upstream preparation to evaluate the material and the ability of the supply chain to adapt. In terms of deconstruction, the environmental and economic viability must be assessed. Sustainability and level of circularity mainly depend on the material characteristics of the building to be deconstructed as well as on local secondary resource markets. Early resource assessment and diagnoses must also be conducted to dismantle, disassemble, and sort the materials as thoroughly as possible. For this, assessment experts must be trained and qualified.

The circular economy described above is based on the efficiency and optimisation of the use of resources and the reduction of waste throughout the life cycle of goods and products while creating economic opportunities. Because of its ecological and socioeconomic impacts, the construction sector is considered to have a high potential to generate value and take advantage of practices on a variety of scales.

Based on these principles, Augiseau (2020) defines the circular built environment as a

> *built environment designed in a modular and flexible manner, sourcing healthy materials that improve the quality of life of the residents and minimise the use of virgin material. It will be built using efficient construction techniques and will be highly utilised owing to shared, flexible, and modular office spaces and housing. Components of buildings will be maintained and renewed when needed.*

When applying a circular economy in the built environment, given the interdisciplinarity context and the large number of constituents involved, the number of complex practical questions increases. At the end of its life, it is essential for a construction project to maintain an efficient waste management perspective and, for circularity, to maximise the reintegration of materials and products in the circular cycle and to minimise outgoing resources from becoming waste. This requires knowledge of the actual given resources and their values (e.g. their nature, quantity, and state) to initiate and apply procedures to give the outgoing materials a secondary or cascading (i.e. material reused in different loops with different functions) prolonged life.

The six elements of ReSOLVE mentioned above can be applied to products, buildings, neighbourhoods, cities, regions, or even to entire economies. Below, actions are described and applied to the built environment.

Regenerate

Regenerating and restoring natural capital is applied when regeneration allows for efficient and circular building performance by reducing the negative impact and use of primary resources aimed at protecting, restoring, and increasing the resilience of ecosystems. When using net zero strategies, low-impact design, materials, and operations are promoted, and environmental and economic impacts are reduced. Renewable biological resources can be extracted and reused via anaerobic digestion, composting, or biorefining, thereby generating energy and cutting carbon emissions, as well as allowing biological material to be returned to the Earth's soil by replenishment. Resilience can enable services and assets to be reliable, robust, and adaptable during stress in progress and sudden shocks. Resilience can be enhanced through chosen design and materials, including flexibility, redundancy, disassembly, and reusability.

Share

Optimising the use of assets by sharing can allow the construction sector to use spaces, infrastructure, and vehicles more efficiently. Ridesharing, which enables users to rent out and share vehicles (increasingly electric), is already well established and affords more use of a smaller number of assets, which reduces impacts such as traffic congestion, carbon emissions, and air pollution. Thus, in a built environment, it is possible to rent out or share underused spaces, building materials, and equipment.

Co-location, shared and flexible working spaces, and more people working from home (especially in the aftermath of COVID-19), connected by the Internet, are increasing both in densely populated cities and in the countryside. When minimising the time an asset is idle and occupying less space, fewer resources are needed to deliver the same function or service. This includes buildings where more people within a smaller footprint make more efficient use of offices and workplaces throughout the 24-hour cycle and sharing vehicles. Sharing spaces can include renting underused spaces for meetings, team away-days, private parties, conferences, and workshops. An online platform can help services to reach a wide audience. Thus, costs for conventional events and space booking agents can be cut, making it simpler for asset owners to take advantage of underused space.

Shared ownership, sharing platforms, and the 24-hour economy enable a shift in the way space and services are used and accessed. This approach can provide savings in costs and additional revenue for owners and operators.

Co-living is also gaining increasing popularity, providing low-cost homes with private bedroom spaces and flexible shared kitchens, dining rooms, and libraries. Sharing economy businesses often use online platforms, or apps to collect and share real-time data and maximise the use of assets. In the built environment, there is an open source of Design platforms, allowing designers to share designs with users so that they can customise or even construct buildings themselves.

The reuse of building materials and components through resale or redistribution creates economic, social, and environmental benefits. Sharing built environment services can also stimulate greater collaboration between clients and building operators, technology companies, platform developers, and other industry partners. Collaboration is already common in parts of the built environment sector but connecting more elements of the supply chain is essential to scale up circular economic practices. There are several reuse platforms on which businesses, charities, and individuals can acquire or give away surplus or unwanted stock and equipment in the construction, retail, and office sectors. This supports reuse and new supply chains for secondary materials and encourages assessment and reports concerning the impacts of reusing, upcycling, and recycling of items.

Optimise

To optimise assets, products, and systems in the built environment, they must be operable at maximum efficiency and performance. The key to achieving this is to maintain materials and components at their highest value while adopting design and construction processes to maximise efficiency and simultaneously promoting reuse and repurposing. Digital design methodologies, e.g. building information modelling (BIM)-supported design, can help to optimise performance, while offsite construction with modular components and elements reduces waste produced onsite. To minimise primary material use, components and materials can also be reused to construct new buildings, repurposed for use in infrastructure, or transferred for use in other sectors. Additive manufacturing technologies such as 3D printing can also optimise material use when manufacturing bespoke building elements.

Design with regard to material longevity ensures long-term durability, utilisation, and value of assets. Durable low- or zero-carbon materials can reduce the climate impact and maintenance costs and extend the economic viability of a building. Standardised components manufactured offsite for higher quality control minimise the risk of structural faults and reduce long-term maintenance requirements. Designing for longevity ensures that assets are used optimally throughout their life cycles. Furthermore, designing for the flexible use of buildings enables clients to switch the use of a building at a later date, for example, from commercial to residential.

Reverse logistics is a closed-loop approach that uses remanufacturing, refurbishment, repair, reuse, or recycling to recover and process materials and products after the point of consumption. Incentivised return policies help to drive the flow of materials and products through the supply chain; for example, telecommunication companies often lease cell phones to customers to ensure they are returned. Collaboration between supply chain stakeholders helps to consolidate materials and create the scale needed to build a reverse logistics supply chain.

Loop

To loop implies keeping products and materials in cycles, prioritising inner loops, and cascading. Loop materials and components can occur in both the biological and technical cycles (see above), creating new uses for materials by remanufacturing and recycling machinery and equipment or biodegradables.

Regular maintenance, refurbishment, and repair help maintain materials and products at their maximum utility.

Focusing on the design of the disassembly during the design phase increases the ability of efficient second use and reuse pathways for the components and materials. It also enables a higher integration of recycled materials and components from other industries. Monitoring and tracking the performance of materials and components are also critical to enable looping opportunities further down the line.

Maximising the use of repurposed materials, components, and structures supports circulation within the industry and reduces the need for virgin materials. Remanufacturing maintains materials, components, and structures in use for a longer period. Integrating different construction sites and other industries enables materials and structures to be transformed or repurposed. Thus, coupled with modularity, the disassembly design allows the structure to be changed easily and maintains recourse value.

Virtualise

It is possible to displace resource use with virtual use. A number of digital apps and services have been developed to replace physical marketplaces. They virtually match supply to demand, making it easier to share and exchange goods and services. Digital virtual services can also facilitate real-time maintenance issues that required physical intervention in the past. BIM is a digital tool that can share information relating to all phases of a building's, material's or product's life cycle and can be used throughout the supply chain by designers, contractors, and building operators. During the operation of a building, BIM gathers data to facilitate monitoring processes, enables preventative maintenance, and upgrades and modifies the systems and components. The use of BIM allows multiple stakeholders to collaborate efficiently during the design, construction, and operation of buildings. Furthermore, it enables an optimised design and supports efficient performance and maintenance of buildings and provides opportunities for recycling and remanufacturing.

Video conferencing and virtual meeting rooms enable users to achieve the physical experience of face-to-face meetings via a virtual platform. Participants share information, documents, and presentations via platforms such as ZOOM, Skype, or Microsoft Teams. This was very frequent during the Covid-19 lockdown, e.g. project collaboration in real-time via interconnected colleagues around the world. This development has facilitated flexible working from different workplaces and reduced business travel and its environmental and financial impacts.

Open design and operating standards allow components or systems to be easily exchanged or upgraded. Open standards increase compatibility, support the exchange of knowledge and information on systems performance, and enable faster and more effective upgrades of systems and components that can be shared using open virtual networks.

Exchange

A slow transition from a static approach to products, services, and design is evolving toward a focus on sustainability through optimised, flexible, and user-focused design and operation approaches. This enables increased environmental efficiency and minimises the negative impact. The business models mentioned above, i.e. leasing, performance-based models, and flexible use design, also increase efficiency. Digital technologies promote the pathway of this development, and new approaches have been developed and adopted.

For buildings, replacing fossil energy use with the generation of renewable energy using wind, solar, and heat-pump technology or heat through closed-loop systems (i.e. anaerobic digestion) can be a part of the zero carbon emission strategy and fulfil circularity requirements. Buildings that adopt these strategies can also be connected to the public grid to share renewable energy. This would also promote the balance of the grid, minimise the need for fossil fuel energy generation, and increase efficient decentralised systems, thus, contributing to low or zero carbon emissions.

The use of alternative material inputs and changes in the selection of products and materials used in the construction sector decreases the environmental impact and costs. Materials that are difficult to reuse and recycle can be replaced by materials from renewable sources. Replacement of traditional solutions with innovative and advanced technology solutions with longer life cycles, modular repairs, flexibility in upgrading, and active disassembly are other examples of exchanges in the circular economy.

Another alternative is to replace the product-centric delivery model with new service-centric models. Contracts are based on certain performance delivery outputs or services instead of selling a product and offer incentives after use to return for repair, reuse, or recycling. The model offers a more function-based delivery to a client, where the function, not the item itself, is sold to the client and the ownership alongside installation and maintenance is not the client's asset (e.g. to light a space, the client pays for the lightening of a space, not the light items). This promotes maximum use.

The seven layers

The concept of construction in 'layers' implies that a building comprises separate but interconnected layers, each with a different lifespan. The concepts of the different layers include the following.

- *System* – beyond the scope of the building, such as the district or city, includes the structures and services that facilitate the overall functioning of the system, such as roads, railways, electricity, water and wastewater systems, telecommunications, parks, schools, and digital infrastructure
- *Site* – the fixed location of the building
- *Structure* – the building's skeleton, including foundation and load-bearing elements
- *Skin* – the façade and exteriors

- *Services* – electrical installations, such as wires and equipment, and heating, ventilation, and air conditioning (HVAC) systems, such as pipes, ventilation equipment, and heating systems
- *Space* – the internal fit-out, such as walls and floors
- *Stuff* – the rest of the internal fit-out, such as furniture and lighting

The application of this concept of building layers implies that each element can be easily separated, dismantled, and removed, thus facilitating reuse, remanufacturing, and recycling. For example, façades or heating systems can be designed and fitted as independent entities, integrated with other building systems, but not merged. Constructing separate layers with different lifespans allows each element to be repaired, replaced, moved, or adapted at different times without affecting the building as a whole, thus, reducing unnecessary obsolescence and increasing flexibility of use and longevity over time. Examples of actions that involve integrating the elements of ReSOLVE and the seven layers are suggested by Zimmann et al. (2016) as follows:

- Regenerate
  - *System* – Extraction and reuse of biological resources by anaerobic digestion to create and supply energy to the public grid
  - *Site* – Detoxify and regenerate brownfields to revive the biosphere
  - *Structure* – Nature-based design solutions with low-impact materials
  - *Skin* – Use of green walls and surfaces to integrate with the facade for composting and reuse purposes
  - *Services* – Recycling biological nutrients and producing biogas by anaerobic digestion
  - *Spaces and stuff* – Use of compostable and biodegradable materials
- Share
  - *System* – Own-produced energy back to the grid
  - *Site* – Online applications for space sharing
  - *Structure* – Reuse of structure materials, such as timber and steel beams
  - *Skin* – Pool of shared products, equipment, and personnel
  - *Services* – Reuse of equipment, pipes, metals, electronics, and so on.
  - *Space and stuff* – Maximise space usage
- Optimise
  - *System* – Optimise transport links between neighbourhoods and urban facilities
  - *Site* – Usage of local produced renewable energy sources by the public grid
  - *Structure* – Designing by consideration of life cycles, adaptability, and reusability
  - *Skin* – Leasing of façade performance-based agreements
  - *Services* – Leasing of sensor-based lighting and energy
  - *Space and stuff* – Designing for flexibility and maximised use of daylight and natural ventilation
- Loop
  - *System* – Circular flows of renewables and recourses with usage adaption over time
  - *Site* – Renovate and reuse of existing buildings for alternative use
  - *Structure* – Design for disassembling and regenerate buildings for mixed use
  - *Skin* – Design for modular construction and offsite prefabrication

- o *Services* – Open Design and Operation Standards. Rainwater harvesting, grey-water recycling, and onsite battery storage
  - o *Space and stuff* – Remanufacture products and materials
- Virtualise
  - o *System* – Digital cloud-based storage system. Smart and AI-system to improve integration
  - o *Site* – Open-source design by accessible online platforms
  - o *Structure* – BIM and 3D virtualising to monitor performance and facilitate maintenance
  - o *Skin* – BIM and 3D virtualisation to monitor performance and facilitate maintenance
  - o *Services* – Smart sensors and access to remote operation and maintenance delivery
  - o *Space and stuff* – Virtual meetings and conferences by digital platforms
- Exchange
  - o *System* – Approach of integrated urban circular design
  - o *Site* – Construction with alternative sustainable and low-environmental impact materials
  - o *Structure* – Usage of sustainable and low-environmental impact materials
  - o *Skin* – Integration of bio-facades
  - o *Services* – Shift to sustainable facility management, SFM, contracting with regenerative approach, as described in Chapter 6
  - o *Space and stuff* – Natural lighting and natural ventilation solutions

2.3 Recourse use and material characteristics

The built environment places significant pressure on climate through global greenhouse gas emissions but also through the use of global resources and materials manufactured for buildings. This also represents a major contributor to carbon emissions during a building's life cycle. Global material use is expected to more than double by 2060, and estimations of materials used in the construction sector are expected to increase by one-third. Carbon emissions should also increase considerably due to the increased use of materials. Approximately, half of the global resources extracted from materials are used in the construction sector. Concrete alone is expected to contribute to 12% of global carbon emissions by 2060. The annual consumption of materials by the global construction industry includes high-carbon-content cement, asphalt, brick, plasterboard, stone, glass, steel, and other metals, such as aluminium and copper, all of which are intensely polluting in their extraction, and energy and carbon emissions intensive in their production.

The construction sector also greatly depends on virgin materials. For sand and unbound materials (e.g. crushed stone, crushed slag, or slate), global need requires the extraction of approximately 4 tons of sand per person, mostly from rivers and coastlines. This implies aggravating existing vulnerabilities, environmentally and from a habitat point of view, in coastal regions. Concerning the consumption of wood from vast amounts of forests, only a limited amount of wood from certified forests, indicating sustainable harvest and maintenance, is in use. In addition, significant quantities of other materials are used with environmental impacts disproportionate to the quantities consumed, such as asbestos (very unhealthy), plastics in window frames and architectural profiles, floor and wall coverings, pipes, thermal and electrical insulation, composite materials, advanced fabrics, adhesives, coatings, paints, and varnishes.

Annual material losses in construction and during the demolition work of buildings are increasing, which results in end-of-life materials or waste materials, equivalent to nearly 40% of the mass when extracted. Material loss includes a variety of materials, including concrete, bricks, gypsum, tiles, ceramics, wood, glass, metals, plastics, solvents, asbestos, and excavated soil, and most of these fractions can be reused or recycled. Conversely, metals are universally well-recovered from construction and demolition works due to their high secondary value. Some countries have a high share of recycled mineral materials; however, most are downcycled with a loss of material quality and low secondary value which does not account for the embodied carbon content.

A circular built environment should be based on a model that considers both technique and business to keep materials and resources in use as long as possible (ideally forever) in a closed cascading cycle of extended use, reuse, and recycling, where the potential for reuse of materials is immense. Transforming the linear built environment into a circular built environment offers significant opportunities for the sector to reduce its material and carbon emission footprints. A circular economy, which keeps materials in use at a high value, can, therefore, make an important contribution to achieving multiple sustainable development goals, specifically, SDG 12 (responsible consumption and production), SDG 11 (sustainable cities and communities), SDG 9 (industry, innovation, and infrastructure), and SDG 13 (climate action).

The transition to circularity

Stakeholders at every stage of the built environment value chain play a role in the transition to a circular built environment, but the initial responsibility lies with those who have the greatest capacity for decision-making and who set directions and actions throughout the supply chain. These individuals include policymakers (especially public sector policymakers at all levels of government), investors (both private and public), and clients (both private and public). These stakeholders have control over business models and contractual frameworks in the built environment. Their actions can maximise impacts at all scales along the value chain by setting targets for material use and reuse.

Policymakers

The first step is to identify policy changes that support a circular economic transition. This can include the development of national, regional, and municipal policy frameworks; economic incentives such as VAT reductions on circular economy services and assets; specification of circular public procurement measures; or the convening of partnerships between public and private sector organisations to promote collaboration. Within public planning policy, requirements covering disassembly strategies, requirements of specifications regarding construction materials, proportions of reused and reusable materials, and mandatory lifecycle cost calculations can support circular economy adoption.

Investors

Delivering commercial full-scale construction projects that test the elements of a circular economy approach can promote technical and commercial viability. To promote the delivery of these projects, new valuation techniques are needed, which unlock additional value by eliminating structural waste and promoting latent long-term value. This means that high operational costs and buildings that are inflexible and cannot be deconstructed attract lower valuations than cheaper-to-run, flexible, and de-constructible alternatives. The need is to define the tools and data required to facilitate better valuations over a construction project's entire life cycle and to look beyond purely financial metrics to broader environmental, social, and governance considerations to judge success.

The client

A whole-life value can be used to assess design decisions for all construction projects. The clients instruct the design and construction of their projects with consideration for the entire life cycle, facilitated by contracts that include design, construction, operation, and maintenance. The management of these contractual works is defined in terms of whole-life economic, environmental, and social metrics. The result is a built environment consisting of healthy buildings with long-term values. Clients' assets last longer and have the ability to change their use rapidly, affordably, and frequently compared to the linear economy approach. Greater collaboration can inspire new ways of thinking, working, and delivering value. Opportunities have also been identified in building operations and sharing economies. The first step for construction clients could be to bring all stakeholders together in partnerships to enable more collaborative construction project delivery and facilitate knowledge sharing. Clients could serve as leaders in this respect by setting new expectations for circular procurement and communicating these to the supply chain. This approach can share the outcomes and provide mutual benefits.

Another important step for clients, as well as investors, is to identify the financial opportunities inherent in buildings that are more long-lasting, flexible, and provide health and well-being for users. This could be promoted by better knowledge transfer and increased awareness among clients regarding the value and benefits of circularity and buildings. By working with policymakers through different circular economy collaboration platforms and establishing links through public and private partnerships, clients can have the opportunity to connect with those able to change regulations or introduce incentives to make the systemic shift to a circular economy.

Circular building design

Circular building design requires a significant shift from current design approaches concerning material flows, manufacturing processes and conditions, use and reuse. An expansive and systemic view, as well as a deep understanding of ecological principles, must be adopted. A circular building design can be described as a building designed, planned, constructed, operated, maintained, and deconstructed consistently with circular economy principles. This includes optimising the use of a building throughout its life cycle, integrating the end-of-life phase into the design. Circular design challenges the development of products and materials that minimise the use of primary raw materials. As the name implies, the focus of circular design is to reduce the value loss of the embedded material by maintaining its circulation in closed loops. These loops, such as reuse, repair, remanufacture, refurbishment, or recycling, extend the material life cycle and improve resource productivity. As happens in nature, the material, its parts, or its constituents at the end of their life become a resource, feeding new cycles of use, within or even outside of the original application scope.

The extraction, manufacturing, and distribution of raw materials in the construction supply chain have a significant environmental impact. All materials contain embodied energy and carbon, but many rely on processes that generate other greenhouse gases and have a higher embodied carbon equivalent. Some materials, such as concrete, PVC, MDF-boards, most glues, and paints, also contain additional chemicals. The impact of these can be unknown and the combined 'cocktail' effect is concerning. If materials are thought to be toxic to humans or wildlife, this is referred to as embodied toxicity. This can cause impacts throughout the material's life cycle, i.e. during a manufacturing process, through off-gassing during use by building occupants, or when recycled, remanufactured, or eventually disposed of.

According to Antonini et al. (2020), the circular design qualities are as follows:

*   *Reused* – Using building parts and components already present onsite or re-claimed elsewhere
*   *Recycled* – Seeking building components made of low-value by-products or waste materials
*   *Renewed* – Using materials that are replenished continuously by responsible agriculture and forestry
*   *Compostable* – Choosing materials that can be biologically degraded into natural substances
*   *Durable* – Using components that resist the cascading of use and reuse
*   *Pure* – Favouring components that consist of a single material instead of a blend
*   *Simple* – Choosing low-tech, common solutions rather than complicated ones
*   *Reversible* – Making it possible to undo connections without damaging joined components
*   *Location and site* – Recognising and responsibly developing the qualities of a place

There are quite a few definitions of recycling, but for the built environment, recycling can be defined as recovering and reusing materials at a similar level of value or quality compared to downcycling, where materials are recovered and reused at a lower value or quality level. For example, the term recycling is often applied to materials such as paper; however, in reality, paper is almost always downcycled because of the shortening of its fibres. Many current definitions of recycled content do not define what is in the material, and it is not possible to recycle materials at a similar level of value. The term recycled content can be improved by including factors that describe the value or quality of a product. Are all contents known, especially additives that provide materials such as paper, plastic, and metal-added functional qualities?

Upcycling improves the existing quality of the material for its subsequent reuse. The material can be upcycled under the following conditions.

*   When the current downcycling is improved, the material is recycled at a similar value instead of a lower level. For example, high-grade steel is separated from

motors containing copper contaminants; therefore, the steel can be remelted at the same level instead of being downcycled.
• When a degraded material is repaired for effective reuse, for example, an additive is added to a plastic to repair its damaged molecular strings so that the material can be reused for high-value purposes.

Sometimes, the recycling information includes a statement that the material comes from renewable resources. However, it is important to determine whether and how these materials can be recovered and reused as nutrients. This avoids undermining the renewability designation by increasing the demand vs. supply of a given material. Wood is renewable until it is burned for energy, at which point it rapidly becomes unsustainable. In contrast, 'non-renewable' elements such as silver and gold could satisfy users if effectively recovered and recycled. The recovery and reuse of materials in biological and technical cycles (shown in Figure 2.1) are the main determinants, rather than their designation as renewable or non-renewable. However, as time accelerates product cycles and material demands, the focus is on rapidly renewable materials rather than renewable materials. Sources such as algae can be scaled up for food and non-food production, and the derived products qualify for attention, which can be used in biological or technical cycles, as shown in Figure 2.1.

The circular design choices that satisfy these requirements make it easier to close the material loops.

The main features of the above material characteristics include:

• *Durability* – Because the material can only be recovered if it remains intact during use and disassembly, it must withstand the cascade of repeated recovery and reuse cycles. Moreover, durable and hardwearing materials often age beautifully, thus maintaining or even increasing their value over time. Therefore, selecting the right material at the initial design stage is crucial.
• *Compatibility* – building materials are compatible if they can be easily reused in different assembly stages, reconfigured, and recombined over time. A simple geometry of connection devices, versatile fixing systems, and standardised dimensions are features that maximise material compatibility to its reuse potential.
• *Reversibility* – Even durable materials can only be recovered when they can be removed from their location by technically and economically feasible operations. Reversibility describes the requirement of a material to be easily disconnected from other surrounding building elements. Dry mechanical connections, such as bolts and screws, are, thus, more effective than wet and chemical joints such as mortar or glue. The high reversibility of the fixings not only allows the recovery of the material at the end of its useful life but also facilitates maintenance and repair within the operational stage.

To maximise the recovery of the material for reuse, it is important to carefully deconstruct or dismantle a building or structure. Several approaches can be used to promote material reuse during the design process. Design for deconstruction or adaptability and deconstruction is one approach to close the construction material loops and extend the duration of the structure. Another approach is design for reuse, where reclaimed material is included in the design of a new structure. If the layout is similar to the previous one, this approach can be useful.

However, several design adjustments are required. Design Approach for Manufacturing and Assembly implies that the construction material and products are fully manufactured offsite and assembled onsite. The ability to assemble and disassemble is fundamental to enable the deconstruction and recovery of construction materials and products. Design for disassembly is a principle that is suitable for reuse through minimised, accessible, and reversible connections between the construction material and products. Economic demand, proper disassembly routines, and assembly control performance can be the conditions for this approach. A blended approach to design for deconstruction and reuse implies allowing reuse of a building's parts, repaired or properly disassembled in new applications, thus, prolonging the life cycle of the materials and products. The difference between these two approaches is that the first one involves recycled building materials and, thus, preserves a smaller value. Therefore, it is most valuable when building materials are reused directly or relocated to a new or existing building. Awareness about building changes over time and proper planning are required to rethink current approaches to construction in order to reach the environmental goals of existing buildings. To achieve a circular economy, awareness about building changes over time and proper planning are required to apply these design approaches to construction. Compared to recycled materials, further advantages of the reuse of materials include saving energy and decreasing the embodied carbon.

Materials passports

Nutrient certificates, often referred to as materials passports, are sets of data describing the characteristics of materials and products with value for present use as well as for recovery and reuse. Passports are marketplace mechanisms that encourage product designs, material recovery systems, and supply chains of possession partnerships to improve the quality, value, and security of supply for materials, with the objective of being reused in continuous or closed loops. It is also beneficial for returning materials to biological systems. This is achieved by adding a value dimension to material quality, based on the suitability of materials for present use as well as for recovery and reuse as resources in other products and processes.

The defined recyclable content is an enabler for recycled content regardless of whether it is recycled. If virgin content is not recyclable, then it will pollute recycling streams; thus, recyclability is just as important as recycling. Recycled content that is also recyclable at a similar level of value is the objective of materials passports.

It is also important to distinguish between bio-based materials and biodegradable materials. Many bio-based products, such as biopolymers, are not necessarily biodegradable because they contain additives, such as heavy metals, or are combined with non-biodegradable materials. However, bio-based petroleum products are not biodegradable. In particular, it is important to evaluate bio-based and biodegradable materials in the context of the intended material, for example, if it is intended for the biological or technological pathway (see Figure 2.1).

In the case of recyclable content, the material has an infrastructure in place for recovering content. For biodegradable content, the material can often be decomposed in the available biodegradation facilities. However, many biodegradable materials do not decompose quickly enough in industrial composting facilities and are incinerated. To solve this problem, materials can be defined for industrial composting.

The focus of materials passports is distinct in comparison to that of environmental product declarations (EPD), for which the primary aim is to display the carbon emissions and other environmental impacts of a product. EPDs are often based on lifecycle assessments. Factors such as embodied energy, transport distance, and resource consumption play a primary role in EPDs.

One advantage of material passports over general emissions-trading mechanisms is the connection to traceable materials. When materials are clearly traceable, it is difficult to distort the marketplace by issuing extra certificates, as can occur with emissions trading. In addition, nutrient certificates often quantify what is already given value by the marketplace; thus, there is no need to match it with, for example, emissions because the marketplace mechanisms have already established the value.

Materials passports can be an effective mechanism for transforming whole buildings, their products, and materials into material banks. Buildings are early candidates for the transition to material passports owing to the following reasons:

- *Value* – Materials passports can enhance the value of the building for the client and operational management. Instead of used and outgoing materials becoming waste that is expensive to remove, they become part of the value chain of the building. This can also enhance the property lease and resale values.
- *Tracking* – Buildings involve large material flows that are already quantified in terms of volume, location, and other specifications. Inventory documentation frequently refers to materials used to build, maintain, and operate buildings. These are the transition mechanisms for defining material passports parameters.
- *Chain of possession* – Institutions, such as hospitals, educational facilities, government buildings, large corporations, and airports, usually maintain long-term ownership of their buildings and, thus, have a long-term vested interest in maintaining the value of the buildings. This long-term interest can also provide continuity in the record-keeping of the embedded products and materials.
- *Volumes* – Building structures use large volumes of materials in one place. It is preferable to inventory these volumes in one location, onsite, rather than to trace materials and products purchased in small quantities in many different locations.
- *Local material sourcing* – Many different materials with a high volume in a building tend to be locally sourced, and it is relatively easy to identify the chain of possession and to verify the content of the materials with the material supplier.
- *Time frame for the recovery of value from materials in buildings* – Although buildings are large, long-term repositories for products and materials, realistically, many materials used in the construction and operation processes of buildings have a shorter usage period. For example, the packaging of construction materials and topsoil – removed onsite to make way for construction activities and replaced in the form of landscaping, cleaning products, maintenance equipment, floor coverings, and office equipment – has shorter life cycles. In addition, many different buildings are repurposed after only a few years due to changes in users' needs. When this occurs, interiors are often renovated, in which case, material passports can offer economic benefits for the client and operation management early in the renovation process when materials or products are repurposed or converted to resources for reuse. Resource recovery can occur relatively quickly, or decades later, as discussed below.

- *Early candidate products for materials passports in buildings* – The 'low hanging fruit' for materials passports comprises business-to-business products and materials in buildings, involving a chain of possession where the players are relatively well defined. For example, on the technological side, HVAC equipment and other systems have defined chains of possession for products and materials, from manufacture to transport, sales, use, and disposal. This generally represents nearly half the value of many buildings embedded in products and materials. The equipment is often well catalogued for maintenance purposes. Moveable office equipment, such as photocopiers, also often have defined chains of possession because leasing contracts are part of service agreements. On the biological side, office papers in large institutions offer chain of possession opportunities because they are collected and reprocessed in large volumes. Exterior and interior landscaping is often maintained as part of service agreements, which provide a mechanism for quantifying the nutrient contribution of plants and soil. An important characteristic of many of these inventory systems is that they already exist, with accounting systems that can be modified to track and quantify materials. In many cases, regulatory or certifying requirements, construction sector agreements, or individual contractors already maintain a form of materials passport, often digitally, for materials used for building construction. Less well-defined, but still traceable in large institutions, food waste and sewage can be tracked for organic nutrients.

BIM is an accepted standard in the construction sector and includes BIM objects, which allow designers to include generic or actual products in building designs. BIM objects definitely have the capacity to become material passports; however, a challenge lies in the immense volume of raw data that will add to BIM software. Nonetheless, with the capacity of cloud technics and software development, it is preferable to use these existing systems, perhaps as a new dimension of BIM (see more in Chapter 5, in Section 5.2).

Tools of material selection

Several material selection tools are available to assess chemical content due to dangerous content at the manufacturer level as well as at the building component level. Third-party environmental labelling systems are also available for certain materials, both internationally and regionally. These labels approve a certain maximum content of chemicals affecting humans or the environment, dangerous or hazardous emissions, and use of toxic constituents during manufacturing of the material. (See more about third-party labelling in Chapter 7.) One important detail of selecting building materials is maintaining knowledge of chemical reactions between two or more seemingly harmless materials when they are incorporated into one functional entity.

Some contractors have created a list of materials and components containing specific dangerous chemicals to avoid. When uncertainty prevails, the contractor should use the precautionary and substitution principles as guidelines. If there are requirements for verifying material or other criteria from building environmental certifications during contractual work, knowledge of how to execute and verify these certification criteria is of great importance.

2.4 Reduce – reuse – restore – regenerate

Reuse, repair, renewal, refurbishment, remanufacture, maintenance, and upgrading are life extension strategies for building materials and products (see Figure 2.2). Except for reuse, where an unchanged material passes from one user to another, the material and products return to a circular economy after modifications. When a material or a product is restored, that is, to a like-new state (as in renewal and remanufacture), or to a previously working state (as in repair and maintenance), or to an improved/updated state (as in refurbishment and upgrading), restorative activities can use discarded and second-hand materials or recovered parts and components to prolong the life of the resources. Similarly, cascading and recycling are activities in which materials can be re-employed several times in different products. This requires restorative supply chains that support either closed- or open-loop cycles. Closed-loop implies the logistics and processing of moving unwanted materials and products from one part of the overall value chain to an appropriate point in the original supply chain for the product system. Open-loop implies the logistics and processing of moving unwanted materials and products to organisations that are outside the original supply chain for different types of uses. Restoration can occur in both open- and closed loops if materials and products have sufficient value to remain in the system.

One way to improve the reuse potential of materials is to establish 'cascades' of material use and reuse. Cascades can solve various challenges such as:

*   With many products, such as wood or paper, it is not possible to recycle materials at the same level of value or quality due to the deterioration of components during use and recycling. For example, with paper, fibres are damaged and shortened during each reuse until they eventually become unusable.

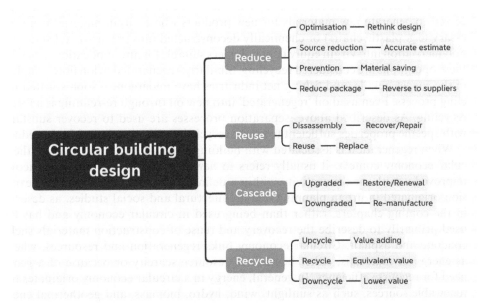

Figure 2.2 Circular building design characteristics (Author original).

- Biomass for paper takes a long time to grow, so if the paper is only used once, and is subsequently incinerated, composted, or downgraded quickly; for example, regarding toilet paper, the use period quickly exceeds the growth period required for replacement.
- Immediate incineration of biomass without using it first in products or after only one use in a product releases carbon emissions far more quickly and in higher concentrations than if the materials are reused repeatedly.

In material cascades, the target is to extend the use of the material for as long as practicable instead of using the material once and then burning or composting it. In a cascade, a material can enter technical loops for multiple uses and then return to biological systems as a nutrient or be incinerated with the ash recovered as a nutrient. Alternatively, if a material such as paper or wood cannot be kept in similar value technical loops, its downcycling can be extended in a controlled way by recovering high-quality fibres for use in products for as long as possible. This type of controlled downcycling is not to be confused with the current form of downcycling, where undefined materials are downcycled with minimal knowledge of their content or pathways.

However, the environmental impacts can vary in open- or closed-loop solutions. In the common language of a circular economy, a product (e.g. a printer or an oven) is regenerated when new parts are substituted for old or defective ones, as in refurbishing, remanufacturing, and upgrading. For refurbishing and remanufacturing, the use of regeneration to describe these activities is synonymous with restoration. Upgrading implies improvement; thus, the term regeneration could be more applicable, but upgrading operations are more likely to be forms of restoration, according to, for example, high user requirements or legislation.

Regeneration is also applicable for the transformation of unwanted materials into new useful materials.

An example is the recovery of plastic bottles, which can then be mechanically processed to generate raw materials for new products where virgin material is not necessary (e.g. plastic chairs) or chemically deconstructed into precursor substances for polymerisation into virgin equivalent polymers suitable for any application. Currently, these operations are considered recycling. Similarly, regenerated nylon fibres employed in the carpet, textile, and fishing net industries have undergone a sophisticated recycling process. Even used oil 'regenerated' into new oil through re-refining is a form of recycling. As described above, separation processes are used to recover substances with specific properties sufficient to create recycled versions of the original product.

When regeneration is associated with buildings, spatial areas, and cities in the circular economy context, it usually refers to new physical forms, uses of structures, improved wealth, or connectivity within a defined geographic location. This expression originated in urban planning and architectural and social studies, as described in the coming chapters, rather than being used in circular economy and has been used primarily to describe the recovery and reuse of construction materials such as concrete and asphalt. Circular economy links regeneration and resources, whether as energy or materials, either because of resource scarcity or because of a general need for recourse efficiency. In general, energy in a circular economy originates from renewable sources, such as sunlight, wind, hydro, biomass, and geothermal energy. Often labelled as regenerative energy, the conditions under which these sources can be considered regenerative are not defined. This can be clarified by the distinction between non-living and living renewable resources, where non-living renewables (sunlight, wind, or rivers) regenerate through a steady input that keeps refilling the resource stocks. Whereas living renewables regenerate through reinforcing feedback, such as more fish means more reproduction and, therefore, more fish. However, these types of renewable resource distinctions are incomplete because they do not address resources such as forests and freshwater aquifers, where the rate of use can far exceed the rate of regeneration and result in irrecoverable depletion. Moreover, even if resources meet these criteria for regeneration, it is not clear how the use of such resources would result in a circular economy that is regenerative in and of itself.

Restorative and regenerative methods have been used to describe the material characteristics of circularity. Restorative refers to a circuit of endless use, reuse, and repair. Regenerative refers to a cycle of life that maintains and upgrades the conditions of ecosystem functionality. These terms can be useful in promoting the concept of a circular economy in the built environment, but can also result in blurring and misunderstanding, as described above.

The use or agreement of definitions for both concepts is lacking, and stakeholders refer to their own interpretations, thereby amplifying the breadth of each term. If the concepts are to have any value for the circular economy in the built environment, similar definitions must be determined for restoration and regeneration.

Morseletto (2020) suggests definitions for restoration and regeneration in the context of a circular economy in a built environment. A simple definition of restoration could be when the material returns to its previous or original state. This is a plain definition that respects etymology and corresponds with most interpretations in the circular economy literature.

Regeneration is not easily differentiated from restoration, but Morseletto (2020) suggests that it promotes the self-renewal capacity of natural systems with the aim of reactivating ecological processes damaged or over-exploited by human action.

However, regeneration is a central principle of a circular economy if it is limited to a single sector of the economy. Regeneration is distinct from restoration; one explanation of how it applies throughout the circular economy is that regeneration is not considered a primary principle, whereas restoration can be a point of reference and a guiding principle for circular solutions. This does not prevent the possibility of other concepts being needed to enhance restoration.

Case 2: Reusable timber structures

Timber construction techniques have a long history of indigenous and vernacular origin up until the present time. However, timber as an urban construction material on a significant scale has mostly been developed from the second half of the 20th century until the present, when housing construction has been the primary use. Currently, in Sweden, for example, 80% of single-family houses are built offsite as a result of a long development history of prefabrication, which has led to a reduction of 20%–25% against the cost of onsite constructed buildings and an 80% saving in time. Although the use of light-frame prefabricated structures has become more common for multistore housing, cross-laminated timber construction is the most widely used.

Timber is a biodegradable, naturally grown, renewable material and is considered a very important construction material for achieving sustainability objectives although its ability to be reused can be an obstacle. As biological materials – sensitive to moisture, mould, and other biological impacts – reclaimed materials demand special care and control if they are to be reused. The benefits of reusing structural timber in construction include the reduction of the building's embedded carbon, prolonging the life cycle of the materials and decreasing the environmental impact. However, timber-based structural construction does not yet entirely follow the principles of the circular economy, nor does it usually consider the entire lifecycle cost of the buildings. Currently, the reuse of structural timber is an obstacle due to the lack of design standards, dismantling practices to prevent damage to the structural timber, lack of a sufficient market for reusable and recoverable materials, building codes and constraints, and fixed dimensions of available new virgin materials, thus negatively impacting the flexibility of design. There are also concerns about the technical performance and safety of reusing structural timber. Glue-laminated timber, common with traditional timber framing, has high reuse potential and offers environmental benefits. These obstacles are mainly the result of concerns regarding costs, inconsistent quality, quantity, and trust.

On average, the highest reuse potential material was prefabricated steel, whereas the lowest was concrete. Nevertheless, the reuse potential of timber is relatively similar to that of steel. When identifying reuse potential, local issues, such as traditional structural systems, local climate, and societal conditions, can be taken into consideration due to the variation from region to region and country to country. When evaluating timber construction, one must consider that timber trusses have a low reuse potential of less than 50%, while timber floorboards have a medium reuse potential of 50% and structural timber has, if properly disassembled, the highest rate of more than 50%. However, it is difficult to dismantle timber components correctly when cleaning, de-nailing, and sizing; moreover, it requires specialist skills and equipment during reclamation and protects timber from exposure to decay. However, if structural timber is connected as a demountable connection with dowels and bolts or holes for wires, its value of reusability can be increased. Of course, it is essential to document this in the material passport described above and, consequently, in a design for dismantling to promote the reuse of timber structures and sections. This can reduce the environmental impact by more than 80%. It is also essential to develop a database of reusable materials and products with unique IDs with connections to BIM tools to promote this design approach for dismantling.

2.5 Regenerative approach to circular economy in the built environment

Regenerative design is based on anticipating the multifunctional evolutions of buildings used in the future, which will be described in the following chapters. In a rapidly changing society, buildings require the ability to adapt to unknown changes and new socio-cultural and demographic issues. It is, therefore, essential to anticipate these changes and integrate strategies that allow the building to adapt to a variety of uses over time. At present, immense quantities of building materials end up in landfills or incinerators long before they have lost any value, quality, or use. To begin, it is essential to define a logical choice in terms of constructive and structural systems, such as columns, beams, and slabs, in order to be able to upgrade the reuse and cascade cycles thereafter. Flexible construction systems can make it easier to disassemble structures and recover, upgrade, modify, or remanufacture building materials. The selection of a flexible construction system allows future users to disassemble a building in its elements, components, and materials to increase the resilience of the building in terms of multifunctionality and flexibility regarding spatiality and use. The modular design of construction systems allows for the reuse of components and materials, thereby increasing the multifunctional capacity of building uses, that is, modular construction systems that allow maximum spatial and user flexibility with parts that can be easily disassembled into reusable building materials and products. Various examples of such an implementation could be in wood, metal, aluminium, concrete, and even in masonry; modular structures (such as containers) or thin steel structures are other examples.

As described in the coming Chapter 6, in Section 6.1, designers must select a flexible construction system that allows for combing architectural elements and regenerative products. A flexible construction system is the key to future modifications by the addition, subtraction, or replacement of the envelope and façade layers. For the chosen system, a space analysis of the building must be performed to evaluate nature-based solutions and connections to ecosystem services. Depending on the microclimate condition of the building site and its geographical location, certain design elements can be more appropriate, thus improving indoor and outdoor air quality, water usage, and increasing the biodiversity, health, and well-being of end-users, enabling cultural and social cohesion, and generating energy.

The final approach of the regenerative design framework is to address building products, optimise the material selection process, and integrate certified products into the building to increase its embedded material value. Each brick, board, piece of wood, or glass in a building has value. Instead of becoming waste, buildings must function as banks of valuable materials and decrease the use of natural resources to meet the resilience capacity of the planet. C2C-certified products or those with similar eco-labels are more beneficial for use because their connection to different cycles generates less of the biosphere or technosphere. Choosing regenerative building products and materials guarantees health, safety, and benefits for humans as well as the environment. These materials or products are designed to be safely reintroduced into biological or technical cycles (as shown in Figure 2.1) and are assembled or manufactured with 100% renewable and non-polluting energy. Regenerative building materials and products are designed to protect and increase clean water resources (as a basis for social and environmental justice). The use of such products also generates chain partnerships to validate each intermediary's recovery and reuse within a manufacturing process. This includes passing a passport for each material and creating a unique database for a building to facilitate reuse for the future. Sustaining the value of materials is the key to the circularity of material use and refining this value is the centre of regenerative buildings. Integrating material passports with reversible building design to optimise circular value chains leads to a significant reduction in resource use. Tracing building materials and products will increase product lifespans and enable product and material reuse, recycling, and recovery. Furthermore, an upgraded cascading approach for recovered materials and products will reduce the generation of waste along product chains in different manufacturing processes and decrease the utilisation of virgin materials, emissions, and depletion of harmful substances.

Bibliography

Acharya, D. et al. (2018) *From Principles to Practices: First Steps towards a Circular Built Environment*, ARUP, Ellen MacArthur Foundation and 3XN Arhcitects. https://www.arup.com/perspectives/publications/research/section/first-steps-towards-a-circular-built-environment, access 2022-08-26.

Antonini, E. et al. (2020) *Reversibility and Durability as Potential Indicators for Circular Building Technologies*, Sustainability 2020, 12, 7659.

Attia, S. (2018) *Regenerative and Positive Impact Architecture-Learning from Case Studies*, SpringerBriefs in Energy, Springer Nature, Cham, Switzerland, 2018.

Augiseau, V. (2020) Circular economy and construction, K. Delchet-Cochet (ed.) in *Circular Economy: From Waste Reduction to Value Creation*, John Wiley & Sons, London, UK, pp. 137–158, 2020.

Bergman, Z. et al. (2018) *The Contribution of UNESCO Chairs toward Achieving the UN Sustainable Development Goals*, Sustainability 2018, 10, 4471.

Ellen MacArthur Foundation (2022) https://ellenmacarthurfoundation.org, access 2022-02-21.

GlobalABC/IEA/UNEP (Global Alliance for Buildings and Construction, International Energy Agency, and the United Nations Environment Programme) (2020) *GlobalABC Roadmap for Buildings and Construction: Towards a Zero-Emission, Efficient and Resilient Buildings and Construction Sector*, IEA, Paris, 2020.

Haliday, S. (2019) *Sustainable Construction*, 2nd edition, Routledge, Abingdon, UK, 2019.

Hansson, B. et al. (2020) *Byggledning – Fastighetsförvaltning (Construction Management – Operation and Maintenance)* (Swedish Only), Studentlitteratur, Lund, Sweden, 2020.

Hart, J. et al. (2019) *Barriers and drivers in a circular economy: the case of the built environment*, 26th CIRP Life Cycle Engineering (LCE) Conference, Science Direct, Procedia CIRP 80, 619–624, 2019.

Lisco, M. et al. (2021) *Challenges Facing Components Reuse in Industrialized Housing: A Literature Review*, International Journal on Environmental Science and Sustainable Development 2021, 6(2), 73–82.

Loftness, V. (ed.) (2020) *Sustainable Built Environment, Encyclopedia of Sustainability Science and Technology Series*, 2nd edition, Springer Science+Business Media, New York, NY, 2020.

Morseletto, P. (2020) *Restorative and Regenerative Exploring the Concepts in the Circular Economy*, Journal of Industrial Ecology 2020, 24, 763–773.

Tirado, R. et al. (2022) *Challenges and Opportunities for Circular Economy Promotion in the Building Sector*, Sustainability 2022, 14, 1569.

UNEP (2020) *2020 Global Status Report for Buildings and Construction: Towards a Zero-emission, Efficient and Resilient Buildings and Construction Sector*, United Nations Environment Programme Nairobi, Kenya, 2020.

Zimmann, R. et al. (2016) *The Circular Economy in the Built Environment*, ARUP, London, UK, 2016.

3   Who – the client and the stakeholders

3.1 The client

The client in a construction project can be defined as an individual or organisation launching and funding a construction project, directly or indirectly. It could be a developer, property owner, or project sponsor (see below) or a public, commercial, or private client with an interest in or control over a construction project. Even in relatively small projects, the client's role is rarely a single-person task. The client's main concern is ensuring that suitable management arrangements are made for the construction project. For the actual project, a competent project management team and the most advantageous and skilled contractor should be selected and appointed. The client must determine the construction project's needs and objectives, what it aims to satisfy and the suitability of the client's overall strategy. Further, the availability of resources must be analysed and determined to successfully develop and deliver the project. This includes giving expression to the project vision, the need for comprehensible and concrete strategies and objectives and understanding and delivering the responsibilities and obligations as a client. When determining the degree of involvement during the development of the project, the client must consider whether external support is required.

Common activities undertaken by clients regarding buildings and construction projects include (not in order):

- Land or property acquisition
- Financing and investment decisions
- Selection and appointment of project management and design teams
- Appointment of a client advisor if appropriate (see below)
- Making decisions regarding design matters for the building
- Making decisions regarding type of contract and procurement of contractor(s) and suppliers
- Making decisions regarding discrepancies between the contract and the contractor's performance during the construction phase
- Negotiating and concluding agreements of lease for the building's users
- Coordinating hand-over to the users
- Initiation and hand-over to the operation organisation
- Long-term planning of maintenance of the building

A client advisor is appropriate when the client or the client's organisation require more knowledge about certain issues, e.g. aside from the tasks above – project business case

DOI: 10.1201/9781003177708-3

development, investment estimations, risk assessment, procurement alternatives, or sustainability and regenerative objectives. A client advisor should understand the project objectives and requirements of the client and provide them with independent and objective advising directly. Furthermore, as a support to the client advisor, an assigned project manager (see Chapter 5) could provide more detailed information from and within the project team, forming the basis for the client's decisions.

As a project sponsor, the client could act as a focal point for key decisions about progress and variations. The project sponsor must possess the skills to lead and manage the client role, have the authority to make day-to-day decisions, and have access to people who make key decisions. Conversely, in the case of an enterprise, the CEO acts as the project sponsor, the executive representative, and the board as the client, the entity that makes the overall decisions. As a client, it is also important to follow the development of the construction market. This can be achieved through statistics and forecasts of future needs. It is also important to understand what forms of tenure exist and what they mean.

The client and sustainability

The sustainability performance and impact of the building and the construction process may be particularly important to the client. The main concerns are meeting the United Nations' Sustainable Development Goals from the base of the triple bottom-line, e.g. using resources effectively, minimising pollution, improving biodiversity, enhancing well-being, and supporting the local community. Resources used during the construction process, e.g. money, energy, water, materials and land, should be used in an appropriate way, without unnecessary waste, poor design, poor quality in construction and supply procedures. All should be affordable, maintainable, and easy to use. When minimising pollution, dependence on materials, energy, and transport must be reduced, especially concerning carbon emissions. To improve biodiversity, use of materials from endangered habitats or species should be avoided and appropriate measures – e.g. restoring wetlands, reforestation, or ensuring a sufficient quantity of planting – should be enacted to enrich existing local natural ecosystems. Further, an environment that promotes and ensures well-being should be facilitated by avoiding inhabitants' exposure to toxic materials, harmful components, and pollutants and enhancing living and sound environments to promote enjoyable leisure, work, and social cohesion. The needs, requirements, and aspirations of local communities and stakeholders can be fostered by involving them in key decisions as the project progresses the in social aspects of the project's sustainability. To successfully manage these issues, the client must consider, identify, and select appropriate targets, tools, and benchmarks as early as the inception phase of the construction process if possible. This is vital in order to deliver a product (e.g. a building) that is as sustainable as possible with minimal resources used in the project budget, thus, making the lowest impact on the environment. (See more about key factors in Chapter 6.)

The client and the regenerative process

Shifting from a sustainability focus to a generative process requires a level of commitment from the client to diverge from the conventional, linear design process management and to reconsider it as an opportunity to learn. A transition from

Learning Level I, doing the same things in a better way (efficiency), to Learning Levels II and III, which generate new levels of systemic understanding (see also Chapter 1), is necessary. This learning process requires the project management team to engage deeply, participate, and be conscious of the Earth and human systems that are essential to the long-term health of the place and community. Thus, the client and project management teams become learning organisations. The client must shift to a regenerative process in the very early inception phase of the construction process through their own initiative or that of the client advisor. As regeneration is a Learning Level III process, the frame of the project objectives must transition from imposing upon nature to a partnership with nature and acknowledgement of the deep relationship between human and Earth systems. The aim is not merely to create a landscape and local habitat for a productive and healthy ecosystem; engagement with the entire system of what makes a place healthy is required. From the perspective of understanding the entire system, an entry point into it can be found through any of its small or large parts. Each is an integral part of a living system, and a key role can be identified for anyone and any system.

Aspects of a regenerative approach

Regenerative processes are established through three essential aspects. Not ordinary steps, they involve continuous improvement over time – the process continually develops as the subject changes. The process must be intentionally sustained long after the construction project is complete. If not, the relationships established during the design phase will become diminished.

The three aspects are:

- Understanding the master pattern and story of the place
- Translating the patterns into design guidelines
- Developing a conceptual design with ongoing feedback – a conscious process of learning and participation through action, reflection, and dialogue

The first aspect requires an understanding of the type of human intentions that the project aims to realise and, largely, the unique character of the place it seeks to inhabit. Contrary to conventional processes, gathering knowledge from different fields, such as water, energy, and soil, can be both fragmenting and misleading. The intentions of the client and other stakeholders must be identified, such as the drivers of the project and what is important to the client and the main stakeholders about the project and the site. This should be determined through a dialogue between the stakeholders expressed in qualitative and process terms.

To learn about the place, considerations should address the status of the ecosystem on site and the project role inside it, how the intentions of the project can support and be supported by the system, how it works and why. One way is to study the historic and present patterns in relation to human and ecosystem relationships, i.e. between humans, plants, animals, hydrology, metrology, and geology. This can be described through general and approximate details – for example, when and why life fully evolved and when, how, and why it changed. Carefully reading the landscape surrounding the project site and developing knowledge about the key intersections where small interventions stimulate the entire system is necessary. The aim is to determine when the investment ensures a greater yield of biodiversity than the actual impact of the project. Thus, to develop the story of the place, other important stakeholders should be engaged in the learning and understanding process of the project's relationship with the complex surrounding ecosystems. The story of a place serves many purposes as a context. First, the ability to sustain and to implement and maintain necessary changes, day after day, with care that comes from a deep connection to place, requires a clear cultural relationship with the specific place. Second, learning about the story of a place enables an understanding of how its ecosystems works and, thus, an understanding of how to achieve a mutually beneficial result. Third, the story of a place provides an integrative context that helps maintain a collective and meaningful purpose. Finally, the story of place provides a framework for a continual learning process that enables humans to interact with their environment.

Once the desired story of a place and its pattern of relationships is defined, the task is to translate it into a conceptual design and design guidelines, which serve as the framework for decisions made in the subsequent phases of the construction process: design and selection of appropriate green materials and technologies, construction, operation, and long-term operation and maintenance. In collaboration with the design team, the client establishes a development concept from the understanding of the story of the place, integrating the project intentions, human needs, and the mutually beneficial relationship with the ecosystems of the site and its surrounding context. The story of a place is married with intentions for the future. This is the point in the regenerative process when the conceptual design can begin.

The core team

To further develop the regenerative process, significant dialogue is required regarding the real issues of the environment and the intentions of the project within the limits of the place. At this point, it is very important to form a core team to maintain these intentions in relation to the place and project. The responsibility of the core team is not about day-to-day activities, but to remember, maintain, and promote the core of the project and to stimulate the resilience of the place and the project. The work of the core team is essential for implementing regenerative processes. Without a core team monitoring the project intentions and understanding the place, the process reverts to old patterns. When the initial design work is finished and the design team disbands, the remaining key stakeholders will need to sustain and develop a conceptual, conscious learning and feedback process in the future. An ongoing core team will ensure long-term continuous monitoring and measurement involving the stakeholders of the place. The team will facilitate the iterative cycles of action, reflection, dialogue, and learning as deepening connections, growing understanding, and mutual caring of the project within the place as it shifts from sustainability to regeneration.

3.2 Management systems

Construction is a complex process with multiple actors, each having different interests in participating in multiple activities during a specific timeframe that require an appropriate level of quality for a specific cost on a given site. The possibility of changing the outcome of the product (the building) during the construction process decreases considerably with time. A project-based organisation undertakes the work, starting from scratch on a new site with new combinations of performers who share a common goal, i.e. the building. This requires a system for managing the process from different perspectives. Most of the active companies in the construction sector utilise some type of management system in their organisation. A management system's objective is to effectively manage the organisation's activity regarding profitability, customer satisfaction, good quality, work safety aspects, and with a minimum environmental impact.

As sustainability issues increase, the overall complexity of a construction project and the ability and knowledge of the project management team must deepen and broaden. This requires the project management team to develop a more strategic and holistic way of addressing these issues. Often, experience gained during the construction process is minimally shared with others in the field, a weakness that reflects the lack of a natural forum for the distribution of information. A system for reporting experience gained needs to be established to enable those engaged in the various stages of new projects to access the knowledge and experience of others, both during and after the project. If no such transfer of knowledge and experience occurs, companies risk losing the opportunity to take advantage of what has been learned and making the same mistakes again. Per International Standard Organization (ISO) 9001 and ISO 14001, which are commonly used in the construction sector, the managerial system is required to assemble information relevant to a project. However, the complexity of the construction process means that special measures are required if the collection of relevant information, including the experience that has been gained and made available to those who require it, is to function properly. Many believe that the ongoing public debate regarding what takes place within the construction sector reflects flaws in the quality assurance system and the lack of a well-functioning system for collecting and distributing knowledge. There is a good reason behind the construction sector's endeavour to identify ways in which the functioning of these two systems can be improved. In temporary organisations, such as those in construction, knowledge of a specific project and the use of routine checklists often play a central role. Knowledge and information often need to be transferred from one actor to the next, similar to athletes in a relay race. At times, it can be difficult for craftsmen to understand from explicit sources how a specific step in the construction process should be conducted to satisfy the demands for sustainability. Tacit knowledge, i.e. doing what one is accustomed to doing without studying drawings or written materials, can play a significant role in such circumstances. The willingness to work in this way (i.e. figure things out on the spot) can be a positive trait, especially when no drawings or descriptions of the exact procedures to be undertaken are available. However, it can also lead to insufficient precision and result in quality, environmental and sustainability requirements not being met.

There are mainly two types of management systems, often combined by companies: quality management systems and environmental management systems (EMS). The first focuses on the intended product, and the latter on the non-intended residuals. ISO 9001 and ISO 14001 stipulate complimentary requirements for such systems. Industry guidelines, e.g. ISO 26000, Corporate Social Responsibility (CSR), and a commitment by UN Global Impact, were established and followed when companies began to focus more on ethical issues. Work safety issues are often covered by national work safety acts or regulations in many countries. Presumably, most well-managed companies already unknowingly meet most of these requirements.

Environmental management systems

The EMS was first systematically implemented in the United Kingdom in the early 1990s. The British standard BS7750 became the basis and starting point followed by the European Eco Management Audit Scheme in the mid-1990s. Beginning in the early 2000s, the leading EMS was ISO 14000, a family of complementing and supporting standards for an EMS. Regarding EMS, ISO 14000 includes the following:

*   *ISO 14001 and 14004* – The actual EMS standard with requirements, and the general guidelines on principles, systems, and support techniques, respectively.
*   *ISO 14002* – Guidelines for using ISO 14001 to address environmental aspects and conditions within an environmental topic area — Part 1: General. This is intended for organisations seeking to systematically manage environmental aspects or respond to the effects of changing environmental conditions within one or more environmental topic areas, based on ISO 14001.
*   *ISO 14005* – Guidelines for a flexible approach to phased implementation. This standard is intended for a phased approach to establish, implement, maintain, and improve an EMS that organisations, including small- and medium-sized enterprises (SMEs), can use to improve their environmental performance. The phased approach provides flexibility that allows organisations to develop their EMS at their own pace, over a number of phases, according to their own circumstances.
*   *ISO 14006* – Guidelines for incorporating eco-design. This is intended to assist organisations in establishing, documenting, implementing, maintaining, and continually improving their management of eco-design as part of an EMS. Organisations that have implemented an EMS in accordance with ISO 14001 are supposed to help integrate eco-design when using other management systems. The guidelines are applicable to any organisation regardless of its type, size, or product(s) provided and are applicable to product-related environmental aspects and activities that an organisation can control and influence, but it does not establish specific environmental performance criteria.
*   *ISO 14007* – Guidelines for determining environmental costs and benefits. This relates to when organisations want to determine the environmental costs and benefits associated with their environmental aspects, e.g. the dependencies of an organisation on the environment, for example, natural resources, and the context in which the organisation operates or is located. Environmental costs and benefits can be expressed quantitatively in both non-monetary and monetary terms, or qualitatively. The standard takes an anthropocentric perspective, that is, it examines changes that affect human well-being, including their concern for, and dependence on, nature and ecosystem services. This includes use and non-use values, as reflected in the concept of total economic value, when environmental costs and benefits are determined in monetary terms.
*   *ISO 14008* – Monetary valuation of environmental impacts and related environmental aspects. This standard specifies a methodological framework for the monetary valuation of environmental impacts (i.e. impacts on human health and the

built and natural environment) and related environmental aspects (i.e. the release and use of natural resources). The referred monetary valuation methods can also be used to better understand organisations' dependencies on the environment. The monetary value to be determined includes some or all values reflected in the concept of the total economic value. An anthropocentric perspective is taken, which asserts that the natural environment has value in so far as it provides well-being to humans. The monetary values referred to are economic values applied in trade-offs between alternative resource allocations and not absolute values.

- *ISO 14009* – Guidelines for incorporating material circulation in design and development. This is intended to assist organisations in establishing, documenting, implementing, maintaining, and continually improving material circulation in their design and development in a systematic manner, using an EMS framework in accordance with ISO 14001. It can also help in integrating material circulation strategies in design and development when using other management systems and can be applied to any organisation regardless of its size or activity. It provides design strategies for material circulation to achieve material efficiency objectives of an organisation by focusing on the type and quantity of materials in products, product lifetime extension and recovery of parts, and materials of products.
- *ISO 14050* – Vocabulary. This standard defines the terms used in the fields of environmental management systems and tools to support sustainable development. These include management systems, auditing, and other types of assessment, communications, footprint studies, greenhouse gas mitigation, and adaptation to climate change.
- *ISO 14051* – Material flow cost accounting – General framework. This standard provides a general framework for material flow cost accounting, traces the flows and stocks of materials within an organisation, quantifies them in physical units (e.g. mass and volume), and evaluates the costs associated with those material flows. It includes common terminology, objectives and principles, fundamental elements, and implementation steps.
- *ISO 14052* – Material flow cost accounting – Guidance for practical implementation in a supply chain. This standard provides guidance for the practical implementation of material flow cost accounting in a supply chain.
- *ISO 14053* – Material flow cost accounting – Guidance for phased implementation in organisations. This is a practical guideline for a phased implementation of material flow cost accounting that organisations and SMEs can use to improve their environmental performance and material efficiency.
- *ISO 14055* – Guidelines for establishing good practices for combatting land degradation and desertification – Part 1: Good practices framework. This is the standard for establishing good practices in land management to prevent or minimise land degradation and desertification. It does not include management of coastal wetlands.

Other standards in the 14000-series include environmental auditing and related environmental investigations (14015, 14016), environmental labelling (14020–14029), environmental performance evaluation (14031–14034, 14063), lifecycle assessment (14040–14049, 14071–14073), and greenhouse gas management and related activities (14064–14069, 14080, 14090–14092).

ISO14001: terms, definitions, and the process

When using environmental management systems overall and ISO 14001 specifically, there are some special terms that are useful for overall sustainability and construction. The most important aspects are as follows.

- *Environmental policy* – Intentions and directions of an organisation related to environmental performance as expressed by its management
- *Environmental aspect* – Elements of an organisation's activities, products, or services that interact or can interact with the environment.
- *Environmental impact* – Change to the environment, whether adverse or beneficial, wholly or partially resulting from an organisation's environmental aspects
- *Environmental objective* – A result to be achieved set by the organisation and consistent with its environmental policy
- *Compliance obligations* – Legal requirements that an organisation must comply with and other requirements that an organisation has to or chooses to comply with
- *Indicator* – Measurable representation of the condition or status of operations, management, or conditions
- *Environmental performance* – Measurable result related to the management of the environmental aspects, i.e. measured against the organisation's environmental policy, environmental objectives, or other criteria, using indicators
- *Prevention of pollution* – Use of processes, practices, techniques, materials, products, services, or energy to avoid, reduce, or control (separately or in combination) the creation, emission, or discharge of any type of pollutant or waste to reduce adverse environmental impacts
- *Continual improvements* – Recurring activity to improve environmental performance
- *Life cycle* – Consecutive and interlinked stages of a product (or service) system, from raw material acquisition or generation from natural resources, design, production, transportation/delivery, use, end-of-life treatment, and final disposal
- *Interested party* – Person or organisation that can affect, be affected by, or perceive itself to be affected by a decision or activity (e.g. customers, communities, suppliers, regulators, non-governmental organisations, investors, and employees)

The process of an EMS of ISO14001 is cyclic with the aim of continual improvement. It follows the concept of the Plan-Do-Check-Act (PDCA) model which provides an iterative process to achieve continual improvements. Briefly, it comprises:

- *Plan* – To establish environmental objectives and processes to deliver results according to the organisation's environmental policy
- *Do* – Implement the planned processes
- *Check* – Monitor and measure the processes against environmental policy and objectives, and report the results
- *Act* – Take actions to continually improve

Imagine the EMS process as a high-jump contest by the company. The actual high-jump equipment is an environmental policy. The ground on which the equipment stands represents the compliance obligations (see above) and the minimum requirement to fulfil the jump; it is impossible to jump underground. The high-jump bar represents environmental objectives. The process requires jumping over the bar without dislodging it (manage to fulfil the environmental objectives), and then raising the bar to a higher level to jump over (continual improvement) until the world record is reached (sustainability status) or beaten (regenerative status).

Implementing and certifying an EMS do not imply that the organisation is equipped to prevent environmental impact; it only implies that the organisation is aware of its environmental aspects and has, by continual improvements, begun to work with preventative measures for its environmental impact.

An environmental policy should contain a summary of the organisation's long-term intentions in a few sentences. The top leadership of the organisation establishes and ensures the policy with commitments of continual improvements, to fulfil its compliance obligations, to be relevant to its activities, to set a framework for its environmental objectives, to be documented and communicated to the organisation, and to be available to the interested parties.

To identify the organisation's environmental aspects, it must go through all its activities and determine all the aspects that have some kind of environmental impact. Further, the organisation must determine the 'worst' cases of environmental impact and establish them as significant environmental aspects. In addition to this work, compliance obligations must be established by identifying all relevant regulations and codes together with the organisation's own voluntary commitments for its activities.

Finally, the environmental objectives are formulated from one or more significant environmental aspect(s) and compliance obligations, considering risks and opportunities. The environmental objectives should be consistent with the environmental policy, measurable (if possible), monitored, communicated, and updated as appropriate.

The next step is to achieve environmental objectives. In this plan, the organisation should determine what will be done, what resources will be required, who will be responsible, when it will be completed, and how the result will be evaluated (measured).

When the EMS is implemented, the organisation should appoint those primarily responsible(s) for its implementation and operation. Further, relevant education of the staff, communication, and documentation of the content of the EMS should be ensured. Routines for operation and procurement with a lifecycle perspective and risk assessment should be established.

During the operation, the performance must be monitored and measured relative to the objectives and routines must be ensured to clarify consequences regarding discrepancies or adverse effects. From a lifecycle perspective, the organisation should ensure the requirements of the product or service during design, construction, procurement, and communication to external providers, such as contractors, suppliers, and end users. To ensure environmental performance, the organisation should audit the process and validate the performance result, either by an internal audit programme or by an external audit process. The latter is necessary for ISO 14001 certification.

The last step in the PDCA cycle is to ensure continual improvement through a top management review and to adjust the environmental objectives if necessary. The review should include an analysis of the audit results, reach environmental performance relative to the environmental objectives, changes in compliance obligations with regard to changes in legislation and codes, the status of research and development, stakeholder and market considerations, risks, emergency incidents, and opportunities. When the review is finished, the top management should handle continual improvement by adjusting the environmental objectives and/or environmental policy.

To achieve the environmental objectives of the EMS of the client's organisation, it is crucial to adapt and formulate them through an environmental programme in construction projects. This programme should be adapted to fit the requirements for every single project as an environmental plan and belongs at the start of procurement of the design team and the contractor(s). This plan demonstrates how to meet the overall environmental programme. The environmental plan should include information about the following:

- How to verify
- When to verify (in a time schedule)
- Who is responsible for verifying – main and partial responsibility?
- How to deal with nonconformity, what routine and template should be used

For the client, a main responsibility for the organisation's EMS and a person responsible for each project should be assigned; it should also be ensured that the staff of the organisation and project teams received adequate education regarding the EMS.

Ethical considerations

In addition to quality and environmental issues of management systems, greater focus has been placed on the ethical questions of organisations over the past few years. Consequently, ISO 26000 of CSR was established to compliment ISO 9001 (quality standard) and ISO 14001. In short, CSR implies the economic, environmental, social, and ethical responsibility for an organisation. The economic part relates to managing an activity with financial benefits and the return of invested capital for shareholders. The environmental part aims to minimise the negative impact on the environment and minimise the depletion of natural resources. Further, the social part considers the organisation's activity and human health and welfare, regardless of whether it concerns employees, suppliers, business partners, or consumers. Lastly, the ethical part addresses how the organisation treats economic, environmental, and social considerations together and how different alternatives have been valued from a moral point of view based on the organisation's and the community's ethical values.

Another example of ethical commitment is to adhere to the ten principles of the UN Global Compact, which considers human rights, labour, the environment, and anti-corruption.

The principles concerning human rights include the following:

1 Businesses should support and respect the protection of internationally proclaimed human rights.
2 Make sure that they are not complicit in human rights abuses.

The principles concerning labour include the following:

3 Businesses should uphold the freedom of association and effective recognition of the right to collective bargaining.

4 The elimination of all forms of forced and compulsory labour.
5 The effective abolition of child labour.
6 The elimination of discrimination in respect of employment and occupation.

The principles concerning the environment include the following:

7 Businesses should support a precautionary approach to environmental challenges.
8 Undertake initiatives to promote greater environmental responsibility.
9 Encourage the development and diffusion of environmentally friendly technologies.

Finally, the principle of anti-corruption stipulates that:

10 Businesses should work against corruption in all its forms, including extortion and bribery.

3.3 The stakeholders

A construction project is cross-disciplinary, thus, multiple factors and stakeholders must be considered to achieve solutions and alternatives, especially when sustainability is considered. Project stakeholders are defined by most of the project handbooks as individuals and organisations who are actively involved in the project, or whose interests may be affected by the execution or success of the project. A stakeholder can be any individual or group with power to be a threat or benefit. The stakeholders in a project can be divided as internal and external. The internal stakeholders are members of the project group or those who finance the project. External stakeholders are those affected by the project in a significant way. An important part of the management of the project and the EMS is to identify and manage different stakeholders affected by the project and determine how they will react to project decisions.

A stakeholder analysis is used to identify the stakeholders and their claims on the project are essential to form and choose strategies in the stakeholder management process. When identifying each stakeholder, it is not sufficient to focus only on formal structures of project organisation. Informal and indirect relationships between stakeholder groups and their importance must also be assessed. To effectively manage stakeholder interests, it is not enough to identify their demands and needs. Project management must also identify the relative power that different stakeholders have on the implementation of the project. Project managers must work collaboratively with the stakeholders to clearly identify which roles are appropriate for each stakeholder throughout the development of the construction project by using the stakeholder power/interest matrix (see Figure 3.1) and conducting an analysis of responses to the following questions:

- How interested is each stakeholder group in impressing its expectations on the project's decisions?
- Do they mean to do so?
- Do they have the power to do so?

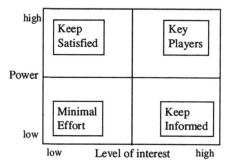

Figure 3.1 The stakeholder power/interest matrix (Persson, U. and Olander, S. (2004)).

Other approaches can also facilitate stakeholder engagement, including SWOT (strengths, weakness, opportunity, and threat) analyses which facilitate collaboration with stakeholders to better understand what will drive the success of sustainability in construction and threats to be aware of.

The stakeholder management process consists of planning, organising, motivating, directing, and controlling the resources used to accommodate stakeholder strategies and to analyse the impact that different stakeholders could have on the success or failure of the project. The stakeholder management process should ensure a more effective project management process due to increased knowledge of the potential impacts (positive and negative) that different stakeholders have on the project's outcomes. Experience from completed construction projects increasingly entail stakeholders' impacts, especially from external stakeholders who are not directly involved in the project, such as affected neighbours, the general public and different local and national authorities. From a sustainability point of view, the aim should be to conduct the project so that as many stakeholders' needs as possible can be fulfilled.

Various powerful stakeholders can use or provide economic instruments to promote sustainability. From this analysis, the external demands on the project, general conditions, internal demands, and specific conditions can be specified. The general conditions of the organisation include the application-related conditions directly linked to the organisation and its way of working. One main condition is the organisation's environmental policy and environmental management system, including relevant parts of environmental objectives. Another main condition is the social and cultural impact of external stakeholders' influence. The third main condition is the economic conditions for the organisation (business-related) and external stakeholders' influence.

Engagement with stakeholders

Engagement with stakeholders across sectors offers the opportunity to gain feedback from a variety of perspectives, especially those that will support the implementation of zero carbon roadmaps (especially across the private sector). Multi-stakeholder engagement also creates strong community buy-ins to maintain momentum through leadership transitions. Stakeholders to be engaged are listed below; it must be noted that some of these cover different sectors that may not efficiently coordinate or collaborate:

- National governments design and implement policies that enable or disable the uptake of sustainability in buildings and construction. National governments act as regulators and can play an important role in facilitating partnerships among other stakeholders.
- Subnational governments play a critical role in the development, implementation, and enforcement of policies. In addition to their regulatory role, cities and states can convene actors across sectors and act as owners of public buildings.
- Utility companies have significant building data and valuable relationships with clients and tenants that already include payment and financing. In some cases, utilities must comply with legislation to reduce their emissions. Utilities can, therefore, be either significant barriers or enablers to action on sustainable buildings and construction.
- Property developers make decisions about how property will be used, including cost–benefit assessments for different construction approaches. These early decisions can have a great impact on the options that are considered in a construction project.
- Financial institutions provide mechanisms to make the necessary upfront investments for sustainability in construction, with repayment often coming from the energy-saving benefits that develop over several years.
- Architects and engineers and professionals who lead technical project design determine what is possible within the parameters set by clients or developers. Professional and educational training provides experts with the knowledge they need to develop sustainability in construction designs. When expanding the role and responsibility of building design professionals to a regenerative development, it also demands that they become increasingly clever to promote the regenerative development's ability to create projects that convince clients to expand the scope of work beyond a narrower set of functional performance requirements.
- Manufacturers and suppliers, companies that make equipment and systems, determine what products are available on the market, and whether building upgrade solutions are sold with a system view or more gradually replacements over time.
- The construction professional workforce must interpret project designs and bring them to life, and there are many risks for real-life installations to fall short of the sustainability envisioned in the designs on paper. Professional training is critical for the construction workforce to achieve sustainability in construction projects with the required level of quality.
- Users and occupants are responsible for the buildings' energy service consumption, for paying for any building upgrades, paying for energy bills, and benefiting from improved energy services.
- Civil society organisations, such as consumer and environmental advocates, or social service providers, can provide analytical capacity and grounded expertise to inform all stakeholders and improve government decision-making. Civil society can represent the perspectives of communities, building occupants and users that may otherwise be absent from buildings and construction dialogues.

Multiple stakeholder engagement processes enable government decision-makers to explore and assess the relevance and feasibility of different approaches, taking into consideration various needs and perspectives. Stakeholder engagement can also build relationships with key players, drive policy acceptance, and improve participation and compliance.

While there are currently no real formal quantitative ways of 'measuring' the success of regenerative development, proponents highlight a more qualitative approach in the establishment of a shared vision, responsibility, and ownership among a broader range of relevant stakeholders by being more involved as co-creators and providing greater assurance to maintain the initial ambitions of a project. In addition, we consider the feedback necessary for continual improvements to the client's EMS. The regenerative contributions over the next decade are not likely to be initially judged in terms of the number of built projects and their associated carbon reductions although it is critically important that regenerative projects demonstrate that they can achieve substantial project numbers and reductions.

Key actions of stakeholders

A set of stakeholder actions is key to enabling a successful implementation to meet sustainability and regenerative development in construction projects, such as capacity building, financing, and multi-stakeholder engagement. Capacity building enables people to understand and act on information that can support the achievement of zero emission, efficient and resilient buildings, and regenerative development. Financing is critical for turning policy and project ideas into reality. Multi-stakeholder engagement incorporates feedback from implementers and those affected, builds trust, and creates strong community buy-in to maintain momentum through leadership transitions. The following key actions are important for capacity building and finance.

- More efforts are needed to collect and centralise data management, to understand how best to use it, and how to balance data privacy and transparency. Data sharing can be made obligatory or incentivised if it is voluntary.
- More university education and workforce professional training – availability of a university curriculum for sustainability in construction is currently low. Workforce training programmes are more common, especially for the construction workforce on topics such as green building materials and energy efficiency.
- More awareness is needed of integrated approaches with urban development and connections with air pollution and public health.
- Institutional coordination across multiple stakeholders. Coordination is critical in several dimensions:
  - Across government institutions, especially to avoid duplication of initiatives.
  - Across organisations supporting governments, especially to improve ease of government participation in initiatives.
  - Between public, private, and civil society sectors to enable continuity and access to centralised information.
- Encourage the formation of global and regional green building communities. Regional and national green building councils and national alliances are examples of communities with convening power capable of bringing together actors across the fragmented value chain.

- Financing options are available, but access is often difficult due to a lack of information, awareness, and standardisation. Increased awareness can enable incentives to raise industry ambition and interest.
- Policies, in general, need to address the decision-making processes in design and construction in a way that addresses the entire carbon life cycle, stimulating full decarbonisation of the sector.

Capacity building, a knowledge transfer

Capacity building is a knowledge transfer used to increase awareness, access, and analysis of data and information. This includes data and tools to assess building emissions and energy consumption, information about co-ordination across institutions in the public sector or across sectors, and awareness of green buildings in education and training curricula. Capacity building activities can increase overall awareness across all relevant stakeholder groups, improve the decision-making process, and encourage more choices of sustainability. Training professionals who work in real life with the built environment is essential to increase resources and capacity to deliver carbon zero emission, efficient and resilient buildings.

Capacity building includes increasing awareness within government institutions on the benefits of green buildings and construction, such as economic impacts, public health and well-being, and benefits to the energy sector and the environment. Shared goals and coordination within relevant government institutions and NGOs can enable an improved policy context. For example, national policy can create an enabling environment for local governments to accelerate action towards green buildings, and local policies are required for strong implementation.

Capacity building as training programmes for service and product providers for buildings and construction (architects, developers, contractors, vendors, etc.) and building owners increase awareness of green building policies, programmes, or incentives for sustainability in construction. This increases professionals' ability and willingness to implement these programmes. Educational programmes, including primary, secondary, vocational, university, and adult education, enable increased knowledge of green buildings. Certification or accreditation for professionals in the building sector can motivate more people to participate in educational training programmes and increase awareness of who is trained to support green buildings and construction. Training of product and material manufacturers can inform them of how to comply with product and building standards as well as capacity building to enable the development and deployment of low-carbon solutions, i.e. increasing efficiency in manufacturing and construction processes and design, employing circular design principles, strategies to increase recycling and reuse, training and access to tools for financiers and clients to better identify, assess and finance investment opportunities in the zero carbon efficient and resilient building sector. Particularly important is a better understanding and assessment of the benefits of zero carbon, efficient, and resilient buildings within the broader context of climate risk exposure of buildings as assets. Moreover, capacity building is necessary to create stakeholder networks among policymakers, clients, and financiers to set up project pipelines. Information to the general public regarding increasing awareness, improving decision-making, and promoting more sustainable choices is also included in this knowledge transfer. This includes information about benchmarking programmes, certification programmes, building passports, labels, educational resources, and information on utility and government programmes.

Regarding data management, data on building stock and typologies, energy consumption, and emissions is a critical first step in understanding the starting point, the baseline. From this starting point, developing ways to make improvements towards zero emission buildings and calculate the multiple benefits from the decarbonisation of buildings, positive impacts can be made.

Finance

Zero-emission, resilient, and efficient buildings and construction often face barriers to financing because they require upfront investments for benefits that develop over several years. Incentives and financing encourage building and construction stakeholders to make decisions regarding the support of green buildings. Financial tools particularly relevant to financing clean energy for buildings may include the following:

- *Urban development funds* – Dedicated funding for urban development projects, which can be directed towards sustainable urban development projects.
- *Infrastructure funds* – Dedicated funding for infrastructure projects, which can be directed towards sustainable infrastructure projects.
- *Dedicated credit lines* – Funding delivered through banks for specific purposes, such as sustainable buildings or development projects.
- *Risk-sharing loan/loan guarantee* – Large organisations, such as governments, international banks, or aid organisations, cover the risk of payment default to allow banks to fund a project with lower costs and better loan terms.
- *Green bonds* – Bonds that can be used to bundle funding associated with sustainability in projects.
- *Preferential tax* – Direct funding from the government to reduce or eliminate taxes for sustainable products and services.
- *Grants and rebates* – Direct funding provided by the government, an organisation, or programme during or after the purchase of a sustainable product or service.
- *Energy performance/energy service contracts* – Contracts for services or delivered savings that are typically delivered by an ESCO and can include a range of energy efficiency services and products.
- *Procurement purchase and lease* –The purchase or lease of sustainable products and services. Leasing enables the use of energy-efficient products on a rental basis to reduce capital expenditure.
- *On-bill/tax repayment* – An approach where any recurring bill, such as utility bills, insurance bills, or home improvement store bills, can collect small amounts of money over a long period of time to pay for energy efficiency purchases in smaller payments. An extension of on-bill finance, tax repayment is where the tax authority uses recurring tax payments as a means of collecting money over time. The most common of these is PACE finance, which uses low-interest-loan repayments on the property tax bill until the purchase is fully paid.
- *Community finance and crowdfunding* – Collective funding from a large number of people connected either locally or through a call for funding.

Bibliography

CIOB (Chartered Institute of Building) (2014) *Code of Practice for Project Management for Construction and Development*, 5th edition, Chartered Institute of Building, John Wiley & Sons, London, UK, July 2014.

Cole, R.J. (2020) *Navigating Climate Change: Rethinking the Role of Buildings*, Sustainability 2020, 12, 9527.

EN ISO 14001 (2015) *Environmental Management Systems – Requirements with Guidance for Use*, CEN-CENELEC Management Centre, Brussels, Belgium, 2015.

GlobalABC/IEA/UNEP (2020) *GlobalABC Roadmap for Buildings and Construction: Towards a Zero-emission, Efficient and Resilient Buildings and Construction Sector*, Global Alliance for Buildings and Construction, International Energy Agency, and the United Nations Environment Programme, IEA, Paris, 2020.

Haliday, S. (2019) *Sustainable Construction*, 2nd edition, Routledge, New York, NY, 2019.

Hansson, B. et al. (2017) *Byggledning - Produktion (Construction Management - Construction Phase)* (Swedish only), Studentlitteratur, Lund, Sweden, 2017.

ISO (2018) *Discovering ISO 26000:2010, Guidance on Social Responsibility*, International Organization for Standardization, Geneva, Switzerland, 2018.

ISO (2021) *Standards by ISO/TC207/SC01*, https://www.iso.org/committee/54818/x/catalogue/p/1/u/0/w/0/d/0, access 2021-04-21.

Persson, U. and Olander, S. (2004) *Methods to estimate stakeholder views of sustainability for construction projects*, Proceedings, Plea2004, The 21th Conference on Passive and Low Energy Architecture, Eindhoven, The Netherlands.

Reed, B. (2007) *Forum: Shifting from 'Sustainability' to Regeneration*, Building Research & Information 2007, 35(6), 674–680.

UN Global Compact (2021) *The ten principles of UN Global Compact*, https://www.unglobalcompact.org/what-is-gc/mission/principles, access 2021-04-25.

4   Where – the location

4.1 The location

The uniqueness of a construction project can entail two aspects: the facility (the product) and the site where the facility or building is situated. The site or location of a building is a key variable in design and management decisions. When the building addresses aspects such as energy use, indoor climate, and material productivity, its location contends with aspects at both local and regional levels. Examples of local aspects include urban microclimates, accessibility to neighbourhood buildings, security, and local biodiversity. On the regional level, aspects include community demands, transportation systems, air quality, public health, and emergency preparedness. Most of the assessment tools on the market have focused on the product, while very few have assessed a specific site. In accordance with a conventional environmental impact assessment (EIA), an EIA can be modified into a simplified triple bottom-line sustainability assessment of a construction project site, that is, an EIA-led integrated assessment. This simplified site assessment is easy to include in a sustainability programme or as a baseline for the regenerative development of a construction project. It is fundamental to assess a construction project site and its surroundings because it influences construction project performance, and vice versa. Such a sustainability assessment (SA) could contain the following elements, organised according to the different aspects of the triple bottom-line.

Environmental aspects

The first part of the triple bottom-line is divided into two subsubsections, the first of which concerns aspects regarding *geology and hydrology*, and could include:

*   *Topography* – The ways in which the surrounding landscape impacts the site
*   *Type of soil* – The impact of the nature and condition of the ground on the site should be considered when deciding on a foundation strategy for the building
*   *Wet areas* – Risk assessment of increased water penetration, damp conditions, flooding, and so on.
*   *Polluted ground areas* – High risk of health impacts and possible demands and costs of decontamination
*   *Water areas* – Conditions for recreation opportunities, biodiversity, rich flora and fauna, and opportunities for eco-cycling
*   *Wetland areas* – Conditions for recreation opportunities, biodiversity, rich flora and fauna, and opportunities for eco-cycling

DOI: 10.1201/9781003177708-4

- *Ground water level* – Ground impact on the site (could be conditional on the choice of foundation strategy), risk of depressions on site and surrounding areas, conditions for recreation, enriching flora and fauna, and opportunities for eco-cycling

Concerning *flora and fauna* on and surrounding the site, the SA could include:

- *Biotopes* – Occurrence of trees, plants, and diversity of animals worth protecting (in general and for the actual site)
- *Biodiversity* – Conditions for preservation of biological variations of trees, plants, and animals
- *Panhandles and swathes* – Conditions for continuous green panhandles to support plant and animal biodiversity

Aspects of *climate and air quality* include:

- *Normal variation* – Climate data on an annual, seasonal, and daily basis
- *Wind and shelter* – Conditions affecting the orientation of a building or facility and the baseline of its energy system by normal and deviating circumstances of predominant wind and where places for shelter could be found. Specifically, future increasing impacts due to climate change must be considered.
- *Sun and shadow* – Conditions affecting the orientation of a building or facility and the baseline of its energy system by normal and deviating circumstances of predominant directions of direct solar radiation and identifying the effects of shadow.
- *Precipitation* – Conditions affecting the envelope and foundation of a building or facility. This includes normal precipitation during a year, and calculated heavy precipitation during a year, decade, and centennial. Specifically, future increasing impacts due to climate change must be considered.
- *Damp and dry conditions* – Affect a building's construction and energy system. Specifically, future increasing impacts due to climate change must be considered.
- *Cold and warm conditions* – Affect a building's construction and energy system. Specifically, future increasing impacts due to climate change must be considered.
- *Air pollution* – Conditions affecting a building's indoor climate. The risk of increases in local air pollution due to future climate change must be considered.
- *Climate change* – Effects on the local climate depending on global climate change regarding:
  o Precipitation changes in intensity, occurrence of downfall, and amount per event
  o Mean water level changes, changes in water level extremities, and occurrences and changes in chemical content of recipients
  o Mean temperature variation and changes in extremities and occurrences
  o Humidity variations due to seasonal changes, changes in extremities and occurrences, and changes in mean level
  o Wind variations in mean wind direction, occurrences of extreme winds, and forecasts of most likely directions
  o Weather variations in general patterns of weather, occurrences, and predictions of extreme weather situations

- *Climate change* – Impact of local climate variation on plants and wildlife regarding:

  o Biodiversity, changes in the content and quantity of flora and fauna in local biotopes
  o Migration, immigration of new species and displacement or extinction of existing species

Economic aspects

The economic aspects of the triple bottom-line, in terms of *cultural impacts*, are as follows:

- *Land use*
  o Historical perspective of land use
  o Archaeology: identifying possible remains of human activity, determining the responsibility of the client/landowner to perform inventories or excavations or by obligation of financial contribution
  o Current use of the site
- *Surrounding buildings and facilities*
  o Presence, extent, and ages of surrounding buildings for housing
  o Presence, extent, and type of surrounding industries and the risk of disturbing pollution and noise that may occur around the site
  o Presence, extent, type, and ages of other surrounding facilities and the risk of disturbing pollution and noise that may occur around the site
- *Local traditions* – Of buildings and construction, including special characteristics, construction and materials, local workforce specialty, locally produced materials, and construction products.
- *Conveyances* – Former important roads, paths, and sea lanes for the transportation of goods and humans

The next part of economic aspects includes *existing infrastructure*:

- *Heating and air conditioning* – Existing systems of central heating and air conditioning supply
- *Sewage and drainage* – Existing systems of sewage and drainage, local or municipal operated
- *Recipient bodies of water* – Water areas for buffering sewage and drainage
- *Body of water* – What recipient bodies of water does the site's environment drain into
- *Existing roads and streets*
- *Waste* – How local public waste disposal is arranged
- *The electricity grid* – Local suppliers and locations of connections to the grid include:
  o Broadband grid: Local operators and how to connect to the grid
  o Telephone network: Local operators and existing network
  o Cell phones: Local operators, their network, and actual coverage
  o Broadcasts and telecasts: Local and regional networks and the operators

Further on, various economic aspects concerning *existing conveyance* include:

- *Traffic systems* – Typical traffic, traffic security measures, and traffic intensity surrounding the site
- *Transportation in the site surroundings* – Typical existing transport and intensity of transport of hazardous materials

Lastly, various *market aspects* of the site include:

- *Market values* – Of the surrounding properties
- *Values of taxation* – Whether they differ from the market value of the surrounding properties
- *Demand and supply* – Of similar properties in the neighbourhood of the site
- *Level of rental agreements of commercial properties* – In the neighbourhood of the site
- *Level of rental agreements of similar types of flats* – In the neighbourhood of the site

Social and cultural aspects

The third part of the triple bottom-line comprises *social and cultural* aspects regarding qualitative measurements, as follows:

- *Territory* – Prerequisites for private, semi-private, semi-public, and public zones
- *Security* – Information concerning the types and frequency of the most common crimes perpetrated in the neighbourhood
- *Well-being* – Factors concerning the turnover of tenants and commercial premises in the neighbourhood
- *Comprehension* – How the design of the surroundings can be explained, the ease of orientation, and the ease of understanding, operating, and maintaining technical systems of importance
- *Accessibility* – The level of accessibility for disabled persons to the actual site, and generally in the neighbourhood
- *Flexibility* – The needs of flexible solutions of importance
- *Groups* – The basis of division of groups in the neighbourhood
- *Common facilities* – The selection of facilities for common activities, and meeting points in the neighbourhood
- *Service* – The selection of service facilities in the neighbourhood
- *Urban life* – The prevalence of elements including culture, sport, amusement, commerce, and nature in the region
- *Participation* – Social participation linked to the project's stakeholders and by the surrounding community
- *Influences* – Users and future users' requests for influence on the performance of the project
- *Structure of ownership* – The state of ownership of the surrounding real estate
- *Aesthetics*
  - o Design concepts within the neighbourhood
  - o The formation of different outdoor spaces in the neighbourhood
  - o The main aesthetic layout of the neighbourhood

A similar way to classify the site is to adopt an assessment described by the landscape planner McHarg. This method consists of the following steps:

*   Identify and establish the boundaries of the area
*   Assess the site mapping the following:
    o   Climate
    o   Geology
    o   Geography
    o   Hydrology
    o   Soil
    o   Vegetation
    o   Wildlife habitats
    o   Land use
*   Analyse the data assessed within a chosen value system
*   Suggest a solution for the best-proposed land use

4.2 Urban ecosystem services

Ecosystem services are defined by the benefits that ecosystem functions afford humans or by the direct or indirect value they provide to human well-being (e.g. protection from storm surges or heat waves, air quality regulation, food, and fresh water). Ecosystem services are divided into four major categories: provisioning, regulating, supporting and habitat, and cultural and amenity services (see Figure 4.1). Provisioning includes goods that are obtained directly from ecosystems, such as genetic resources, food, fresh water, wood, fibres, and medicines. Regulating comprises all the benefits obtained from regulation by ecosystem processes, such as the regulation of climate, water purification, pollination, erosion control, and some human diseases. Cultural and amenity services include the intangible benefits people obtain from ecosystems, such as spiritual enrichment, cognitive development, reflection, recreation, tourism, and aesthetic experience, as well as their role in supporting knowledge systems, social relations, and aesthetic values. These three categories are supported by the fourth category of ecosystem services – supporting and habitat – which includes all ecological functions necessary for the production of other ecosystem services categories. Examples include biomass production, nutrient cycling, water cycling, provisioning of habitat for species, and maintenance of genetic pools and evolutionary processes.

Different habitats provide different types of ecosystem services, thus, a detailed classification of ecosystem services needs to be adapted to specific types of ecosystems. Urban ecosystems cover areas where the built infrastructure encompasses a large proportion of the land surface, or areas in which people live at high densities. However, according to the ecological interpretation of urban systems, they must also include less densely populated areas due to the mutual influences between densely and sporadically populated areas. Urban ecosystems are especially important in providing services that directly affect human health. However, which ecosystem services at a given scale are most relevant varies greatly depending on the environmental and socio-economic characteristics of each geographic location.

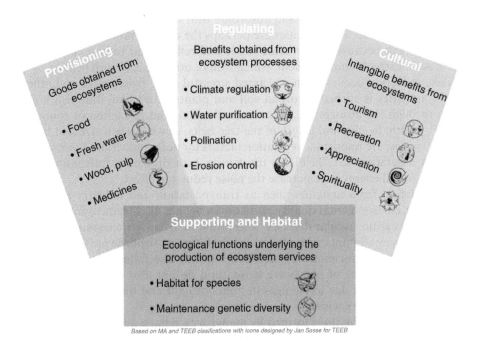

Figure 4.1 Categories of urban ecosystem services (Gómez-Baggethun, E. et al (2013)).

Urban provisioning ecosystem services

The important ecosystem services in urban areas, as defined above, concern provisioning services and food and water supply. Urban food supply is derived mostly from fields adjacent to urban areas and from rooftops, backyards, and community gardens. Regarding water supply, the service aims to secure water to meet human and societal needs. Ecosystems provide fresh water for drinking and other uses for humans by securing the storage and controlled release of water flows. Vegetation cover and forests in city catchment areas regulate the quantity of available water. However, watersheds are often threatened by urban development in and around city water sources; in some places, cities face unplanned urbanisation pressures where serious impacts on water quality and biodiversity could occur, especially around large cities and megacities.

Urban regulating ecosystem services

Urban temperature regulation, noise reduction, air purification, moderation of climate extremes, runoff mitigation, waste treatment, global climate regulation, pollination, pest regulation, and seed dispersers are the most important urban ecosystem regulating services. Regulation of urban temperature occurs when the ecological infrastructure is able to regulate temperature within cities such as urban heat islands. Water areas absorb heat in summer and release it in winter; vegetation reduces heat during the hottest months by shading and absorbing heat from the air through evaporation during low humidity. Trees in cities are perhaps the most significant of the urban ecosystem services because they provide shade and reflect solar radiation in local areas comprised of streets and pavement, which would, otherwise, absorb heat.

Noise reduction occurs when unwanted stressful noise from traffic and other city-related activities is curbed by urban soil and plants. These can diminish noise pollution through the absorption, deviation, reflection, and refraction of sound waves. Density, width, height, and length of the tree belts, as well as leaf size and branching characteristics, are important vegetation factors for noise reduction. For example, the wider the vegetation belt, the higher the density, providing more foliage and branches to reduce sound energy and increase the noise reduction effect.

Air pollution from activities such as transportation and industry pose a major problem for environmental quality and human health, as it leads to increases in respiratory and cardiovascular diseases. Vegetation in urban systems improves air quality by removing pollutants, as trees and shrubs filter out airborne particulates through their leaves. The performance of pollution removal also follows daily variations. At night, the small pores of plants are closed and do not absorb pollutants. Monthly variations also occur due to changes in hours of light and the shedding of leaves by deciduous trees during winter.

Ecological infrastructures formed by mangroves, deltas, and coral reefs can act as natural barriers that buffer cities from extreme climate events, including storms, heat waves, floods, hurricanes, and tsunamis. This type of infrastructure can drastically reduce the damage caused to coastal cities. Vegetation also stabilises the ground and reduces the likelihood of landslides. Climate change is increasing the frequency and intensity of environmental extremes, signalling increasing adaptation challenges for cities, especially for those located in coastal areas. Risk management and vulnerability reduction in cities, based on the combined use of built infrastructures (e.g. levees) and ecological infrastructures (e.g. the protective role of vegetation), are examples of proactive actions to reduce the impact of extreme climate events.

The vast amount of impermeable surface materials in urban areas leads to a large volume of water runoff and, thus, a higher risk of water flooding. Vegetation reduces surface runoff following precipitation events by delaying water runoff through leaves and stems. The underlying soil also reduces runoff by acting as a sponge by storing water in its pore spaces until it percolates through the soil. Urban landscapes with 50%–90% impermeable cover can entail 40%–83% of rainfall to surface runoff compared to 13% in vegetated landscapes. Interception of rainfall by tree canopies alongside streets slows down flooding effects, and green areas reduce pressure on urban drainage systems by percolating water. Other means of reducing urban stormwater runoff include linear features such as bioswales (channels designed to concentrate and direct stormwater runoff while removing debris and pollution), green roofs, and biofilters such as rain gardens or bioretention filters. Green roofs can retain 25%–100% of rainfall, depending on the rooting depth, roof slope, and amount of rainfall. They may also delay the timing of peak runoff, thus reducing stress on storm-sewer systems. Biofilters, such as rain gardens or bioretention filters, reduce surface runoff by creating soil filters that collect, filter, and treat moderate amounts of stormwater runoff using planting soil beds, gravel under-drained beds, and vegetation.

Ecosystems can filter and decompose organic waste from urban sewage by storing and recycling waste through dilution, assimilation, and chemical recompositing. Wetlands can act as filters to reduce the levels of nutrients and pollution in urban wastewater. Similarly, plants in urban soil can play an important role in the decomposition of many types of labile litter. In urban streams, nutrient retention can be increased by adding coarse woody debris, constructing in-channel gravel beds, and increasing the width of vegetation buffer zones and tree cover.

Climate change impacts may worsen in cities because urban areas contain artificial surfaces and high levels of fossil fuel combustion. Urban trees serve as vessels for storing surplus carbon as biomass during photosynthesis. Because the amount of carbon stored is proportional to the biomass of the trees, increasing the number of trees can decrease the accumulation of atmospheric carbon in urban areas. Therefore, tree-planting programmes in cities are an attractive option for mitigating climate change. The amount of carbon stored and retained by urban vegetation is often quite substantial, and urban soils can also act as carbon pools. However, compared to overall city emissions, the amount of carbon a city can neutralise locally through its ecological infrastructure is modest.

Pollination, pest regulation, and seed dispersal are important processes in the diversity of urban ecosystems and can play a critical role in their long-term durability. However, pollinators, pest regulators, and seed dispersers are threatened by habitat loss and fragmentation due to ongoing urban development and expansion. Community gardens as plots of land made available for individual, non-commercial gardening, private gardens, and other urban green spaces are important source areas. In addition, numerous formal and informal management practices in community gardens, cemeteries, and city parks promote the functional groups of insects that enhance pollination and bird communities, which, in turn, enhance seed dispersal. To manage these urban services sustainably over time, a deeper understanding of how they operate and depend on biodiversity is crucial.

Urban cultural ecosystem services

The recreational aspect of urban ecosystems is among the highest valued ecosystem services in cities because the city environment can become very stressful for its inhabitants. Parks, forests, lakes, and rivers provide diverse possibilities for recreation, thereby increasing human health and well-being. The recreational value of parks may depend on characteristics such as biological and structural diversity, but also on built infrastructure, such as the availability of sports facilities and benches. Recreational opportunities may also vary in terms of accessibility, penetrability, safety, privacy, and comfort. Further, various factors may cause sensory disturbance, i.e. if green areas are perceived as ugly, trashy, or too loud recreational value decreases. Urban ecosystems, such as community gardens, also offer multiple opportunities for spontaneous leisure and meetings, an important condition for the recreation of urban commons.

Urban ecosystems are an important provider of aesthetic and psychological benefits that enrich human life with meaning and emotions; for example, aesthetic benefits from green spaces have been associated with reduced stress and increased physical and mental health. A view through a window looking out at greenspaces could accelerate recovery from surgeries, and the proximity of an individual's home to green spaces can be linked with fewer stress-related health problems and a higher general health perception. People often choose where to live in cities, based, in part, on the characteristics of nearby natural landscapes.

Multiple opportunities for cognitive development occur when people experience nature and green spaces, which provide increased potential for stewardship of the environment and stronger recognition of ecosystem services. Urban forests and community gardens are often used for environmental education purposes and to encourage cognitive coupling with seasons and ecological dynamics. In addition, community gardens, cemeteries, and other green spaces have been found to retain important bodies of local ecological knowledge and to compensate for observed losses of knowledge. The benefits of preserving local ecological knowledge include increased resilience in urban systems and sustaining and increasing other ecosystem services.

Place values refer to individual emotional attachment and strong bonds to places, plots, and surrounding garden areas. Having a deep sense for specific places can be a major driver for environmental stewardship. Attachment to green spaces in cities can also lead to other important community benefits, such as social cohesion, promotion of shared interests, and neighbourhood participation. Upgrading the role of urban green spaces provides opportunities for interaction between individuals and groups that promote social cohesion and can reduce criminality. Moreover, urban ecosystems can play a role in defining identity and a sense of community. The latter could suggest that an understanding of how communities are formed enables the design of buildings that will be better maintained and provide better use of surrounding green areas.

Urban habitat ecosystem services

By providing refuge for many species of birds, amphibians, bees, and butterflies, urban ecosystems can play a significant role in biodiversity. Appropriate designs of green roofs can provide habitats for species affected by changes in urban land use. In cold and rainy areas, urban golf courses can contribute to wetland fauna support. Old hardwood deciduous trees are an important resource for species with high dispersal capacity. The diversity of species often peaks at intermediate levels of urbanisation, a level at which many native and non-native species thrive; however, diversity commonly declines as urbanisation intensifies.

Urban ecosystem disservices

Urban ecosystems not only produce ecosystem services as positive benefits and values for humans but also produce functions that are not beneficial to human well-being, i.e. ecosystem disservices. For example, common city trees and bush species can emit volatile organic compounds such as isoprene, monoterpenes, ethane, acetaldehyde, formaldehyde, acetic acid, and formic acid, all of which can indirectly contribute to urban smog and ozone problems. Urban biodiversity can cause damage to physical infrastructures through processes such as microbial activity, which causes decomposition of wood structures; further, bird excrement causes corrosion of stone buildings and statues. The root systems of vegetation often cause substantial damage by breaking up pavement, and some animals are often perceived as uneasy as they dig nesting holes. Green-roof runoff may contain higher concentrations of nutrient pollutants, such as nitrogen and phosphorus, than precipitation input. Other disservices from urban ecosystems include health problems from pollinating plants that cause allergic reactions, dark green areas that are perceived as unsafe, diseases transmitted by animals (e.g. migratory birds carrying avian influenza, dogs carrying rabies), and blockage of views by trees. As some plants and animals are perceived as ecosystem services, animals such as rats, wasps, mosquitoes, and plants such as stinging nettles, are perceived by many as disservices.

4.3 Ecosystem services analysis process

Ecosystem services analysis (ESA) could be a sufficient method for understanding and incorporating ecosystem services into the design of the built environment. An ESA could be executed by the construction project design team in the early stages of the design process as well as by urban planners, ecologists, and policymakers on local or regional levels. The target is to determine the measurable ecological regeneration in an urban or construction site context. The process of developing an ESA into a regenerative design is divided into three steps, as shown in Figure 4.2.

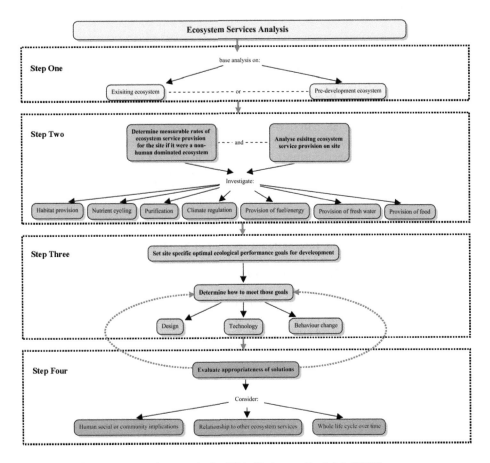

Figure 4.2 Ecosystem services analysis, ESA (Pedersen Zari, M. (2017)).

The first step is to determine whether there is a local ecosystem that can be studied. If not, the base of the design targets could be derived from an ecosystem that existed before the development of the site. Thus, measurable units of provisioning ecosystem services that exist (or previously existed) on the site can be calculated (step 2), e.g. specific data such as annual rainfall and water retention related to the ecosystem distribution of fresh water. Of course, there are gaps in knowledge when measuring different ecosystem services for the chosen ecosystem; thus, each ecosystem contains measurable units to establish initial site-specific targets. In step 3, the optimum environmental performance for the building site is determined by comparing the ecosystem case in step 2 and the performance of the original ecosystem. A mimicking approach can also define the criteria and targets of suitable ecosystem services. An analysis of how the current provisioning ecosystem services are compared with the determined optimum rates should be conducted. The final step is to verify that the suggested targets of ecosystem services and the ecosystem performance of the building site do not negatively impact the performance of other ecosystem services; for example, green facades could increase the ecological footprint by using inappropriate plants for the region/ecosystems, water source for the plants, or overall climate context. It is important that each site should be analysed individually because of different end-uses, local ecosystem contexts, and ecosystem services.

Bibliography

Ali, W. et al. (2021) *Common Approaches of Nitrogen Removal in Bioretention System*, Sustainability 2021, 13, 257.

Gómez-Baggethun, E. and Barton, D.N. (2013) *Classifying and Valuing Ecosystem Services for Urban Planning*, Ecological Economics 2013, 86, 235–245.

Gómez-Baggethun, E. et al. (2013) *Urban ecosystem services*, T. Elmqvist et al. (eds.) in Urbanization, Biodiversity and Ecosystem Services: Challenges and Opportunities Edition: First Chapter: 11 Publisher: Springer Netherlands, pp. 175–251.

Hansson, B. et al. (2017) *Byggledning – Produktion (Construction Management – Construction Phase)* (Swedish only), Studentlitteratur, Lund, Sweden, 2017.

Hansson, B. et al. (2021) *Byggledning-Fastighetsförvaltning* (Swedish only), Studentlitteratur, Lund, Sweden 2021.

Pedersen Zari, M. (2015) *Ecosystem Services Analysis: Mimicking Ecosystem Services for Regenerative Urban Design*, International Journal of Sustainable Built Environment, 2015, 4, 145–157.

Pedersen Zari, M. (2017) *Biomimetic Urban Design: Ecosystem Service Provision of Water and Energy*, Buildings 2017, 7, 21.

Persson, U. (2009) *Management of sustainability in construction works*, PhD thesis, Division of Construction Management, Lund University, Lund, Sweden, 2009.

Roös, P.B. (2021) *Regenerative-Adaptive Design for Sustainable Development A Pattern Language Approach Sustainable Development Goals Series*, Springer Nature, Cham, Switzerland AG, 2021.

5   When – the project organisation

5.1 Project management

To develop a specific product as a building, it is almost always organised as a construction project, depending on the product's (the building) often unique design and site in a time-limited process. To manage such a process, several concerns must be addressed by the client to complete the project to the desired outcome. This section describes how to manage a construction project sufficiently through project management.

The project

A project is a performance conducted over a specific period with a specified start and end date, with the endeavour to create, through its objectives, a unique product, service, or result. Because projects have a specific beginning and end, they are temporary in nature. A project can reach its end when the planned unique product has been created by fulfilling the project objectives or has been terminated before the completion for whatever reason. The temporary nature of a project, in general, does not mean a short duration or does not refer to the product it creates. The project outcome is a unique product – a tangible, quantifiable artefact that is either the end item or a component of it, or a service (in fact, a capability to perform a service), or the result from a process such as research.

From start to finish, a project goes through an entire life cycle, including defining the project objectives, planning the performance to achieve the objectives, doing the work, monitoring and controlling the progress, and closing the project after a successful outcome, the final product. The five stages of a project life cycle or process group are described below:

- The initiating stage – Defines and authorises the project. The client chooses the project manager, and the project is, through a project charter, formulated and described with the purpose of the project, the final product description, the project objectives and requirements, and the overall milestone schedule. The main purpose of this stage is to align the project objectives with the client's business needs and strategy and with the stakeholders' expectations and needs.
- The planning stage – Is when the project manager, together with the project team, develops the project management plan (PMP), including the project scope and refined milestones with activities to meet the objectives. The PMP comprises a collection plan and other documents that define the actions required to achieve

DOI: 10.1201/9781003177708-5

the objectives and requirements of the project, such as scope, schedule, and cost plans. The PMP serves as a base for monitoring how the project is performing and compares the actual project performance results with the intended results.

- The executing stage – Entails managing the implementation of the PMP and the execution of the project as planned. All activities are coordinated to achieve the objectives and meet the project requirements. Approved changes, recommendations, and discrepancies are also managed during this stage. The processes managed involve the coordination of resources, including budget, team members, and time used to perform the project activities, integration and management of activities being performed and ensuring the implementation of the project scope and approved changes. Most of the available project resources are consumed during this stage.

- The monitoring and controlling stage – Ensure that the project stays on the planned track. The project must be monitored and controlled throughout its life cycle, including during the execution stage. The purpose is to ensure that the project performance is occurring as planned, and that corrective action is taken if it goes off track. To fulfil this commitment, the actual project performance data must be collected and analysed in comparison to the intended performance in the PMP; if discrepancies are identified, preventive and corrective actions must be taken to accomplish adequate changes and to ensure that client approval is followed by implementation. Change requests may also derive from stakeholders and should be properly evaluated and processed as described above.

- The closing stage – Verifies that all the required project processes are complete, the product meets the requirements, the objectives are fulfilled, all project-related contracts are closed, the outcome of the project is turned over to the client and its organisation, and, finally, the project team is disbanded. The closing stage also includes conducting a project review of the lessons learned. Terminated projects or projects cancelled before completion should also go through the closing stage.

The project manager

A project manager (another term used is project coordinator, see below) must have the necessary skills and competencies to manage all aspects of a project from inception to occupation. The individual could be a member of the client's organisation or an external consultant. The project manager, representing and acting on behalf of the client, must select, correlate, integrate, and manage different disciplines and expertise with the aim of satisfying the objectives of the project from inception to completion. The service provided must be independent, cost-effective, and to the client's satisfaction and interests at all times.

The term project coordinator should be applied when the responsibility of project management includes only parts of the project, for example, the pre-construction stages. In addition, a clear distinction must be made between the terms when the project manager appoints other consultants, as project managers, or the client appoints other consultants, as project coordinators.

The project manager should be appointed by the inception stage in order to advise the client, through professional knowledge and understanding, and to accomplish the first important steps of a construction project in an optimum manner. In this stage, the earliest and most important in the process, the project manager is also involved in the process of establishing the objectives and targets of the project. This ensures competent management coordination, monitoring, and control of the project to its satisfactory completion, in accordance with the client's objectives. However, depending on the nature and type of the project and the client's in-house expertise, the project manager could be appointed as late as the start of the strategy stage; however, this could deprive the client of important background information and is, therefore, not generally recommended.

The project scope could be defined as developing and progressing the project within the given and agreed boundaries established by the client, i.e. the project objectives. These include requirements on programs for the delivery of the project, the budget, and the requirements for the finished building in terms of function, quality, and any specific requirements regarding performance, e.g. environmental, sustainability, or regenerative.

Understanding the client's needs as clearly as possible and formulating the project objectives is fundamental to project success.

The key role is to motivate, manage, and coordinate project teams. Skilfulness with the tools and techniques of project management is an advantage, but it will not compensate for shortcomings in this key role of project management. Paying respect and professionally recognising the other disciplines in the project team, society, and the environment is also an important obligation of the project manager. This person is to be appointed by the client to maintain full responsibility for the project, a close association with the project team, and an executive role throughout the project with appropriate input from the client. Regardless of the specific duties, depending on the client's expertise and requirements, the type of project, timing of appointment, and continuous duty of exercising control of project time, cost, and performance will always be the focus. An example of the typical terms of engagement for a project manager regarding a construction project is outlined below. These terms can be modified depending on the client's objectives, type of construction project, and other requirements (adapted from Code of Practice for Project Management for Construction and Development, Briefing note 1.02; see list of references at the end of the chapter).

Typical terms of engagement for a project manager during a construction project:

- Analysis of the client's objectives, requirements, and involvement in the project budget framework
- Formulation of how to achieve the objectives within the budget, including quality preferences
- Generally, throughout the project, keeping the client informed on progress and issues, design/budgeting/construction variations, and other relevant matters
- If required, suggesting recommendations to the client regarding:
  o How to select consultants and their terms and conditions of engagement
  o The appointment of contractors and/or subcontractors, including advice about the most suitable forms of tender and contract
- Preparation for the client's approval of the following items:
  o The overall project schedule includes site acquisition, overall planning, pre-design, design, construction, and handover stages

- o Proposals for architectural and engineering services; monitoring progress and taking appropriate actions on submissions concerned with planning approvals and code requirements
  - o The project budget and relevant cash flow affecting the project development
- Finalisation of the client's objectives and confirming with the consultants. Providing the client with data on surveys, site investigations, relevant stakeholders, adverse rights or restrictions, and site accessibility constraints
- Securing the client's approval for any modifications to agreed objectives, approved designs, schedules, and/or budgets from reviews involving the design team and other consultants
- Formulate the management structure and defining:
  - o Responsibilities, duties, and organisation of reporting, for all parties
  - o A quality system with procedures for efficient communication, instructions, drawings, certificates, time schedules, valuations, submission of reports, and relevant documentary returns
- Formulate the tendering strategy
- Inform and, if necessary, advise the client about:
  - o The progress of the design and required drawings/information and tender documents with an approach to optimise costs in construction methods and maintenance requirements
  - o The quality of tender documents
  - o Preliminary construction schedule for the main contractors. Any revisions must fully meet the client's requirements and be conveyed to the project team for action
  - o The progress of the project, especially concerning the budget frame, quality level, and fulfilment of the objectives
  - o The contractual activities concerning the client, including approval/decision points
- Establish cost monitoring and reporting, including feedback to consultants and the client on budget status and cash flow
- Organising and/or participating in reviews with the client and meetings with the project team and others involved in the project, acting as chair or secretary to ensure that:
  - o Information/data is adequately supplied to all concerned
  - o Progress is in accordance with the time schedule
  - o Costs are within the budget frame
  - o Required standards and specifications are achieved
  - o Contractors have adequate resources for management, supervision, and quality control of the project
  - o Relevant members of the project team inspect and supervise construction stages, as specified by the contracts
- The responsibility for:
  - o The project handbook
  - o A motivated and commutative project team
  - o Project progress, costs, and quality; taking action to address all supposed and real deviations
  - o Management of time schedule

- o Coordinate the project team's activities and output
- o Monitor project resources against planned levels and take necessary remedial actions
- o Submit reports and reviews to the client, including time sheets and other data on cost and control
- o All processes within the project
- o Approve, within the framework of the project management agreement with the client, required additional work
- o Identify and resolve existing or potential problems, disputes, or conflicts within the project in the best interest of the client
- o Monitor the cash flow of the project and ensure that the accumulated costs are within the budget frame
- o Verify with the project team for extensions of time or that additional payments are required and, accordingly, advise the client if adjustments are necessary
- o Check the consultants' final accounts before payment
- o Ensure inclusion in the contract of as-built and installed drawings, operating and maintenance manuals, and health and safety files, and ensure that adequate arrangements are made for the client's facility management organisation
- o Request the design team, other consultants, and contractors to supply the client with the required operational and maintenance documents
- Appropriate steps are taken to ensure that site contractors and other regular or casual workers observe all the rules, regulations, and practices of safety and fire prevention and protection
- Participate in the final cost reconciliation of the project and taking action as required

Of course, every construction project is unique depending on the type of client, building, requirements from the actual code and legislation, requirements from the coming tenants, site and project objectives. The above terms of engagement for a project manager during a construction project are merely recommendations and not compulsory terms; the terms always depend on the nature of the project.

The project organisation

Projects are organised to perform a set of activities beyond the limits of the client's organisation's normal existing operations. Distinctions between projects and operations can be made by referring to the definition of a project, which is temporary and unique. Operations are mostly ongoing and repetitive. Although both projects and operations have objectives, a project ends when its objectives are met, while an operation continues to meet objectives, possibly a new set of objectives.

Project organisation is temporary and, in principle, an ephemeral company with its own rules, roles, and communication structures. It is an absolutely necessary instrument for regulating the cooperation of the people involved in the project.

Classic project organisation (see Figure 5.1) includes three main roles: the project owner/client/sponsor (different terms are used depending on which project management school is used), the project manager, and the project team, including project collaborators. This organisation is linear with a specific hierarchy regarding responsibility

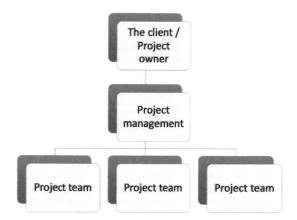

Figure 5.1 Classic linear project organisation (Author original).

and authority. Responsibility concerns occur when individuals in the project organisation perform assignments of the project with or without guidance or authorisation. Authority involves knowing when to handle the right to make decisions, issue orders, and assign resources to meet desired outcomes, granted by the position in the project organisation or by superior authorisation. The role of the project sponsor or the client as a representative includes the following tasks and competencies:

- Sufficient assignment of tasks to the project manager
- Defining project strategies and objectives with the project manager
- Formulating an agreement of organisation framework with the project manager
- Making strategic project decisions
- Making budget framework decisions
- Coordination of the project with all other affected areas
- Support of the project manager when major issues occur

The client representative also acts as a power of attorney for the client's company and could delegate a part or the entirety of this power to the project manager.

On the client side (see Figure 5.1), it is typical for the client to chair the Steering Committee in a project organisation with such a committee (compared with the Board of Directors in an ordinary company organisation). The Steering Committee makes cross-departmental and cross-organisational decisions specific to the project, together with feedback between the project and the client's ordinary organisation.

The project manager and project manager team are the performers of the project, where the project manager's responsibilities and duties are as mentioned in the previous section. The team members' performance includes various tasks, including the implementation of assigned project tasks and maintaining responsibility for the results, contribution of technical expertise, participation, and collaboration in planning, implementation, team meetings, and informing the project manager of the project status.

5.2 Project toolkit or knowledge areas

This section entails the classic steering tools required to manage a project and how, with digital software, to facilitate a construction project to enable its project information to be attached to specific construction parts or materials by using building information models.

The classic project tools

Referring to most of the project management literature, there are various classic project tools to consider; in fact, there are ten tools or processes to use when performing a successful project. Another common expression regarding these tools is the project's knowledge. These are, in general, project constraints divided and categorised into multiple aspects and processes such as integration, scope, time, cost, quality, risk, resources, stakeholders, communication, and procurement. Below are brief summaries of the most important content of the tools.

Project integration occurs when the project is initiated, planned, and executed in pieces using the aforementioned tools or knowledge areas. All these pieces are related to each other and need to be coordinated, i.e. integration management. In general, the integration management tool offers processes for defining, identifying, coordinating, and integrating various activities and processes within each project management process group.

The scope ensures that all the required work, and only the required work, is performed to complete the project. This is what the project should achieve. It involves all the work in delivering the project outcomes and the processes used to produce them. This is the reason for and purpose of the project. Scoping a project comprises the framework of the project and defines what is included and what is not. To perform the work within the project scope, it needs to be scheduled and to identify and assign resources. To manage the project successfully, this work needs to be broken down into manageable tasks. This is accomplished by creating a work breakdown structure (WBS), see Figure 5.2. The WBS is a hierarchical structure that subdivides the project scope deliverables into smaller, manageable tasks called work packages that will be performed by the project team to create the planned deliverables. The structure of the WBS resembles an inverted tree, with work packages at the lowest level of each branch, the leaves. With the work package, the actual necessary work becomes more concrete, detailed, and manageable.

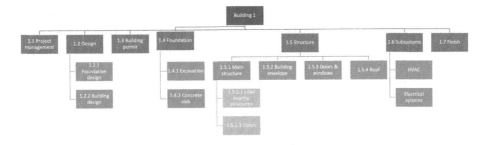

Figure 5.2 An example of WBS for a construction project (Author original).

Time is scheduled to complete the project and is often the most frequent project oversight in project development. This is reflected in missed deadlines and incomplete deliverables. Proper control of the schedule requires the careful identification of tasks to be performed using the WBS and accurate estimations of their duration, the sequence in which they are going to be done, and how people and other resources are to be allocated. However, a work package from the WBS might not be suitable for an individual to perform. A work package can be rearranged or further divided into smaller components, called activities. A schedule entails not only the activities to be performed, but also the sequence in which the activities will be performed, along with a start and finish date. The sequencing of activities is constrained by the dependencies and relations among the activities, and the duration of every activity depends on the available allocated resources. Any schedule should also consider vacations and holidays. Many techniques and software exist to create a time schedule.

Cost entails the budget or budget framework approved by the client for the project, including all necessary expenses needed to deliver the project. Cost is ultimately a limiting constraint for nearly all projects, and few projects can go over the budget without eventually requiring corrective action. The primary tasks of cost management are to estimate and control project costs, and the primary goal is to complete the project within the budget approved by the client. Cost is the value of the inputs that have been (or will be) used to perform an activity or to produce the project outcome. This value is usually measured in monetary units. The budget is an aggregated cost over time for all the resources needed to perform the activities within the project.

Quality is a combination of standards and criteria that the project's outcome, the product, must achieve to perform according to the project objectives. The product must provide the expected functionality, solve the identified problem, and deliver the expected benefit and value. It must also meet other performance requirements, such as sustainability, availability, reliability, and maintainability. The quality of a project is controlled through quality assurance, which is the process of evaluating overall project performance on a regular basis to provide confidence that the project will satisfy the relevant quality standards. This process of managing quality determines the necessary quality requirements and standards to ensure that the planned quality requirements and standards are applied, and to verify that the project and its deliverables meet the quality requirements and conform to the quality standards.

Risk is defined by potential external events that have a negative impact on a project if they occur. During a project, various assumptions and estimations are made, and a few constraints are met. These sources of uncertainty can cause risks that must be managed. A project risk is an event that, if it occurs, has a positive or negative effect on meeting the project objectives. Risk refers to the combination of the probability that an event will occur and the impact on the project if it occurs. If the combination of the probability of the occurrence and the impact on the project is too high, the project manager must identify the potential event as a risk and establish a proactive plan to manage the risk.

Resources are required to perform these activities and can include human resources, material, equipment, facilities, infrastructure, or anything else capable of the definition required for the completion of an activity. The primary purpose is to identify, obtain, secure, and manage the resources needed for the completion of the project.

Before assigning resources to the project, it is necessary to know their availability. Resource availability includes information about what kind of resources are usable, when they are available, and the conditions of their availability. All the allocated project resources together are the basis for estimating activity duration in the time schedule, cost estimation for budget purposes, and planning for procurement conditions. When allocating human resources to the project, it is important to convert activities to roles and responsibilities for teams and individuals, where a role is a set of responsibilities depending on what special competence is needed with or without authority. A responsibility is a piece of work that must be accomplished as part of completing an activity. Authority is assigned to a role that enables the individual to apply project resources, make certain decisions, sign approvals, or accept a completed deliverable. These roles can be managed by using the project's organisational chart to create an organisational breakdown structure (OBS) that lists the necessary project roles with their special responsibilities, competences, and authorities. Then, the OBS is combined with the WBS (see scope above) in an assignment matrix, to create the responsibility assignment matrix, which specifies the relationships between work packages or activities and the roles that perform these activities, defining the human resource requirements for each activity.

Stakeholder management involves identifying project stakeholders and managing and monitoring their engagement in the project. This involves analysing their potential impact on and expectations from the project and project results. Furthermore, it entails the development of a strategy to appropriately obtain the most important stakeholders engaged in the project. See more in Chapter 3.

Communication is essential for project success. It is a critically important component of project management and a common thread that runs through the project life cycle. Planning for communication entails identifying the needs of the project stakeholders, which varies for different stakeholders, and designing a communication approach to meet those needs. Three processes must be completed: plan communication to determine the communication approach, manage communication to make it happen, and monitor communication to ensure that it happened. Historical information and lessons learned are particularly important because they can be used to plan communication wisely based on experience. When establishing a communication plan that describes expectations and needs, various important items should be included. First, the requirements of the stakeholders must be specified, as well as the expected format, content, and level of detail of the information. Second, the senders and receivers should be identified and who is responsible for releasing confidential information regarding the project. Third, the methods used and the frequency of information – daily, weekly, or monthly – should be clarified. Finally, the use of communication technologies such as e-mail, web platforms, chat channels, video conferencing, Facebook groups, and Instagram accounts should be outlined. It is important to plan which tools are needed and who can use different tools. For information that does not change often, written reports are sufficient; however, information that needs to be updated frequently or instantly requires web communication tools.

Procurement occurs when project organisation is unable to perform certain scheduled activities due to a lack of expertise or skills and must outsource these activities. Procurement can be accomplished by acquiring services, items, products, or results from outside of the organisation to complete the project. Procurement management includes processes to plan, conduct, and control procurements. Planning procurements includes making and documenting purchasing decisions on what, when, how, and to what level of quality to acquire from outside the project. Furthermore, potential sellers are identified and a procurement approach is developed. To decide what type of contract will be used for procurement, awareness of the contract types that are supported by the client are important. A contract is a mutually binding agreement between a buyer and a seller that obligates the seller to provide the specified product, service, or result and obligates the buyer to make a payment for it. The procurement plan may include how the make-or-buy decisions will be handled, what types of contracts will be used, what metrics will be used to evaluate potential sellers, requirements for performance, timetable of procurement activities, evaluation criteria for selecting sellers and measuring their performance, and handling legal and currency issues. Although, procurement planning should be undertaken early in the project and may need to be adjusted at any stage of the project as a result of approved changes or other circumstances.

BIM in construction projects

Building information modelling (BIM) enables the sharing of information and data between all stakeholders and participants throughout the project life cycle. It provides a platform for consistent, structured, and perfect data to enable informed smart decision-making at all stages of the project process. BIM is the process of creating information models containing both graphical and non-graphical information in a common data environment (CDE) (a shared repository for digital project information, often cloud-based). The information that is created becomes ever more detailed as a project progresses; when the project is handed over at completion, the client can use the complete dataset during the operation stage, for maintenance purposes, and for deconstruction at the end of the building's life cycle. In other words, BIM is a visual database that allows every team member in a construction project, and in the client's real estate organisation of operation, to centralise data for more efficient collaboration with each other. This makes the construction process and the life cycle of buildings easier, improves building maintenance, reduces cost, and monitors the building's performance.

A BIM-managed construction project can be divided into levels and dimensions. Levels – specifically, maturity levels – facilitate shearing or not shearing the data, defining the degree of collaboration allowed by the software.

- Level 0 refers to the absence of digital collaboration. This means working on a 2D draft and exchanging or distributing it via prints or digital prints.
- Level 1 entails using a 3D model for concept work, but still performing most of the work on 2D. There is no direct collaboration between working on models separately and sharing data via e-mail.
- Level 2 is when all parties use their own 3D model, not necessarily the same one. However, data can be shared very easily because the file format, which can be integrated into other parties' models, is the same. Formats such as Industry Foundation Class (IFC) and Construction Operations Building Information Exchange are the most common. Some countries, such as the United Kingdom, some Scandinavian countries, and various states in the United States, have made it mandatory to use BIM level 2 in public sector projects.

- Level 3 is when a single 3D shared model, called Open BIM, is achieved, in which every party can work and directly add their data. It enables entire lifecycle management of the assets. Open BIM is the process of full collaboration by all parties during all stages of a project, from concept to deconstruction. The idea behind it is to have a single BIM model throughout the life cycle that is accessible to everyone. At this level, it is important to consider all stages, including construction, operation, maintenance, and deconstruction. However, it does not really exist yet (2021), depending on what kind of sheared infrastructure is to be used and the availability (and implementation) of an international set of standards that regulate BIM processes and procedures. Another key requirement is the use of IFC models to describe architectural, engineering, and construction data in a common format. It is an open file format specification that is not controlled by a single vendor and can be easily exchanged between different software without compatibility issues. The IFC model specification is open and available to everyone and is registered by an ISO standard. It is also necessary to use integrated web services as a BIM hub to exchange information so that all parties can access the same information from a single source. Currently, the use of a cloud-based CDE, which is an ISO standard that can be accessed by anyone, anywhere, at any time, is a virtual location where all the project information is stored and accessible to the entire project team. This cloud-based CDE is one of the main challenges for the full implementation of BIM level 3.

BIM dimensions are different from BIM maturity levels. Dimensions refer to the specific way in which specific types of data are linked to an information model. By adding additional dimensions of data, it is possible to gain more understanding and knowledge of the entire construction project, such as how it will be delivered, what it will cost, and how it should be maintained. These dimensions range from 2D to 8D.

- 2D refers to ordinary two-dimensional drawings with height and width.
- 3D entails visualisation and the possibility of using virtual design construction (VDC) applications. VDC technology allows a design and construction team to estimate, coordinate, plan, and build a project in the virtual space long before construction begins. This helps provide the project team members with a clear picture of the project and the best way to proceed with construction, budgeting, scheduling, logistics, and safety (see below).
- 4D refers to the time or construction sequencing. It entails time-related information for a specific element and can include information on lead time, how long it takes to install/construct, the time needed to become operational/harden/cure, the sequence in which components should be installed, and dependencies on other areas of the project.
- 5D entails costs. Accurate cost information can be estimated by the components in the model. Considerations might include investment costs (the costs of purchasing and installing a component), its associated running costs, and the cost of renewal/replacement down the line. These calculations can be made based on the data and the associated information linked to specific components within the model. This information enables cost managers to easily estimate the quantities of a given component on a project by applying rates to those quantities, thereby reaching specific and overall costs.

The 5D dimension detailed above and the dimensions that follow are not uniformly standardised or mutually agreed upon by the construction industry; they can vary between different sources with and within the content. The following are the most common interpretations.

- 6D relates to operation and maintenance. Traditional construction projects have mainly focused on the upfront investment costs of construction. Shifting this focus to a lifecycle perspective, where most money is spent on running costs, should make a base for better project decisions upfront in terms of both cost and long-term material productivity. 6D comprises information to support the client's management and operation of the building to a more effective outcome. These data might include information on the manufacturer of a component, its installation date, required maintenance and maintenance periods, and details of how the item should be configured and operated for optimal performance, energy performance, estimated lifespan, and potentially reuse data. It also enables the client to pre-plan maintenance activities potentially years in advance and to develop spending profiles over the lifetime of the building, estimating when repairs become uneconomical or existing systems inefficient. This planned and proactive approach offers significant benefits compared to a more reactive approach, at least in terms of costs. Ideally, it should be able to continue developing during the operation phase with updates on repairs and updated replacements added in.
- 7D concerns sustainability. The aim is to provide an environmental impact analysis and offer solutions to make a more energy-efficient and carbon-neutral solution. It could also include aspects of regenerative development concerning buildings.
- 8D entails safety, security, and risk issues on the project site, e.g. embedded manuals, precautionary principles, and emergency plans.

When BIM is to be used in a project, it should be implemented immediately from the start. However, shifting a project to BIM might occur at any of the project's stages, but it has consequences in terms of cost, time, resources required, and scale of difficulty. It is also important to establish the driver(s) for BIM. Is it client-driven, or a regulated requirement? Moreover, when ordering a project executed by BIM, it is not recommended to order it solely through these dimensions. It is advisable for clients to order the project so that it is clearly defined in terms of data requirements and information needed, both before and during the project.

A project BIM execution protocol must be established to ensure that BIM is used to maximise advantages and that the entire team is working together in a consistent manner.

Sustainability and regenerative aspects

Sustainability performance and impact may be particularly important to the client. In addition, corporate social responsibility (see the section on Management systems in Chapter 3) could be important in the delivery of the product in the built environment. Sustainability and regenerative requirements could be a part of quality, resource, and procurement management. The requirements could be related to the product itself, the building, and the performance of the product, such as limitations of carbon emissions and energy consumption. In addition, requirements may also be prescribed for any impact on local topography or adjacent areas. Lastly, outcomes in terms of the local community, such as providing employment and training opportunities or the use of local supply chains, may be determined.

From a sustainability and regenerative point of view, the project will be provided with a management framework for the planning and implementation of construction activities in accordance with the client's commitments, the client organisation's EMS system, the project context, the end users, or any other stakeholders. When a regenerative approach is taken, it is important to implement the client's core team (see the section on Core team in Chapter 3) in the project organisation, which will influence key design parameters relating to sustainability, performance, and operational technologies. The overall management criteria for the project is also outlined, including the key success factors for project quality in terms of sustainability management.

Bibliography

CIOB (Chartered Institute of Building) (2014) *Code of Practice for Project Management for Construction and Development*, 5th edition, Chartered Institute of Building, John Wiley & Sons, London, UK, 2014.

Daddey, F. (2021) Project management, https://pressbooks.bccampus.ca/fdaddey/front-matter/introduction/, access 2021-07-08.

Domendos gmbh (2021) https://projectmanagement.guide, access 2021-07-05.

Howard, J. (2015) Benefits of VDC, https://www.mccarthy.com/insights/benefits-vdc, access 2021-07-17.

Joisseaux, B. (2018) The BIM revolution in building management, https://blog.drawbotics.com/2018/11/07/the-bim-revolution-in-building-management/, access 2021-07-16.

McPartland, R. (2014) BIM Levels explained, https://www.thenbs.com/knowledge/bim-levels-explained, access 2021-07-16.

McPartland, R. (2017) BIM dimensions -3D, 4D, 5D, 6D BIM explained, https://www.thenbs.com/knowledge/bim-dimensions-3d-4d-5d-6d-bim-explained, access 2021-07-16.

Sanghera, P. (2019) *PMP® in Depth: Project Management Professional Certification Study Guide for the PMP® Exam*, 3rd edition, APress Media, Springer, New York, NY, 2019.

Terol, C. (2020) BIM Level 3: Is the industry ready? https://www.globalcad.co.uk/bim-level-3-is-the-industry-ready/, access 2021-07-16.

Wildenauer, A.A. (2020) *Critical Assessment of the Existing Definitions of BIM Dimensions on the Example of Switzerland*, International Journal of Civil engineering and Technology 2020, 11(4), 134–151.

6 How – the project performance

6.1 The design process – key factors

The most important decisions regarding sustainability and regenerative development are made during the design phase of the construction process. The pre-design stage (briefing and inception) is especially important – when additional hours of analysing and adopting client objectives to a long-term, holistic, zero (or beyond zero) carbon construction project targets a sustainable or regenerative product outcome: the building. At a relatively low cost for the client, these additional hours of work are between ten and hundredfold more profitable in the long term, but considerably more expensive to adjust later in the construction process. Its value increases exponentially during the design stage.

To begin, this section presents an overview – a roadmap of necessary key actions in design – to be implemented over the next few decades to address the urgent challenge of global climate change and to ensure sustainability towards regenerative development. It then addresses the benefits of design from a sustainability point of view, followed by how nature-based design solutions can contribute to fulfilling these key actions and various examples of these solutions. The next section introduces various nature-based design principles – defined and described as bioclimatic, biophilic, and biomimetic design approaches – as a path towards fulfilling the roadmap requirements. Furthermore, it explains the connections between nature-based design and ecosystem services. The section that follows suggests how to adapt new buildings to zero carbon emissions and sustainability performance over a timeline spanning until 2050, established by UNEP and other global organisations. Key factors are presented regarding the building envelope, HVAC and appliances, cool buildings, and socio-economic factors of healthy homes in the built environment. Lastly, suggestions regarding regenerative approaches to the design process are presented.

A roadmap for key actions in the design of new buildings

This subsection addresses the reduction of operational energy and carbon emissions in new buildings and a roadmap for implementing these actions, successively with tougher measures, until 2050. To maintain current measures for efficiently managing emissions from new and inefficient buildings, the roadmap should encompass several decades. For full-zero carbon buildings to last the entire life cycle, they require measures to reduce the embodied carbon of materials and to shift energy use to renewable energy. Today, approximately two-thirds of the world's countries have no mandatory

DOI: 10.1201/9781003177708-6

or voluntary codes to stipulate the minimum energy performance requirements of new buildings. Thus, it is crucial to increase the number of national building codes requiring a minimum of net zero by 2030 for new buildings to reach whole life net zero carbon by 2050. This requires several key actions:

- To develop local new construction roadmap strategies, with appropriate measures that use a whole life zero carbon approach, including construction materials and energy usage, and aim to achieve a ready-to-operate at zero carbon level by 2030.
- To implement mandatory building codes that set minimum energy efficiency standards for new buildings. Codes should refer to guidelines for locally adapted bioclimatic design principles (see further in this section) for climate resilience and low-carbon materials.
- Reinforce and improve existing building codes with zero carbon content through a long-term improvement cycle that progresses the performance requirements every three to five years, with the aim of achieving zero carbon codes by 2030.
- Minimise the need for cooling devices for space conditioning. Global cooling is the fastest growing service demand in the world. Focus on using the most efficient cooling systems only where necessary and prioritising the use of passive or natural cooling design to maintain thermal comfort.
- Public construction projects should be prioritised, for example, through policies that ensure that all new public buildings meet the requirements of zero carbon.
- Reduce embodied carbon through low-carbon material selection; reduce operational carbon through improved operation and maintenance routines and the usage of carbon clean energy by renewables.
- Increased Knowledge and Information. Increased knowledge and awareness of the benefits of sustainability and regenerative development in buildings will enable clients and users to make better long-term choices and promote beneficial financing with sustainability requirements.

The key stakeholders for sustainability in new buildings include those who can influence the demand for sustainability and regenerative development in new buildings; those who can deliver the results of zero or beyond zero carbon emissions; and efficient, regenerative, and resilient buildings. Additional stakeholders include those who can support the construction process towards sustainability and regenerative development through research, funding, education, knowledge transfer and making appropriate technologies available.

Benefits of designing buildings from a sustainability point of view

When designing new buildings with regard to sustainability, there are multiple and holistic benefits, many of which are linked to the Sustainable Development Goals, especially SDG 7 (affordable and clean energy), SDG 11 (sustainable cities and communities), SDG 12 (responsible consumption and production), and SDG 13 (climate action). Environmental benefits include the reduction of carbon emissions, reduced air pollution with better air quality, reduced use of construction materials, especially non-renewals and non-recyclables, and increased life expectancy of used materials through recourse efficiency. From an energy point of view, a sustainability approach for a building is energy efficient, uses energy from renewable sources, produces energy locally, delivers energy surplus into the public energy grid, and contributes to lowering peak consumption and decreasing possible overuse on the grid by low consumption of external energy. Sustainability in new buildings could also support employment for sustainable services, reduce operational costs, and allocate financial resources available for investment in other parts of the economy. When sustainability endorses good thermal comfort, good lighting availability, indoor air quality, and a pleasant acoustic environment, the physical and mental health, well-being, and productivity of users, employees, inhabitants and students will increase. Furthermore, the sustainability of a built environment will strengthen the value of the property and positively affect the value of nearby properties and the attraction for investments.

An integrated design focusing on sustainability involves professionals from all disciplines of a construction project, including project managers, architects, structural engineers, HVAC engineers, and electrical engineers. In addition, it involves specialists from expertise disciplines, such as acoustic, fire safety, accessibility, and environmental coordinators. Adaptions of key sustainability matters must be applied during the early stages of the design phase of a construction project process to ensure project cost efficiency. Adaptions of sustainability applied later in the design phase entail increased project costs and more unpredictable consequences on other design factors, i.e. a cascade effect. The design tools of considerations include the use of BIM multidimensional technology, daylight-, thermal-, and energy dynamic simulations and lifecycle analysis (LCA) estimations.

Nature-based design solutions

As the world population grows and new urban development increases, natural habitats and nature become more distant from human communities. This urban development consists of new buildings and creates an even greater environmental impact. If nature-based design solutions are integrated, the lack of a natural environment in urban development and the sustainability of places with a substantial amount of health, resilience, and low-carbon benefits could be solved. According to the UNEP (2020), nature-based solutions can be broadly defined as '...actions addressing challenges through the protection and restoration of natural processes and ecosystems'.

There is wide agreement that nature-based solutions can restore this lack of connection between humans and nature and contribute to benefits from a diversity of ecosystem services (see Chapter 4). Research has shown that using vegetation in and around buildings improves thermal comfort, decreases urban heat island effects, reduces operational energy needs, improves air quality and water management, and improves the liveability of the built environment. This is not a new idea; however, due to the urgent need for transitioning to a low-carbon society, several cities, developers, and clients have begun to focus much more on the recognition and implementation of nature-based solutions.

Numerous examples of nature-based solutions and green infrastructure exist globally, especially in urban environments such as Berlin, which has the incentive of creating a 1,000 green roofs. Another example is Toronto, where a green roof bylaw requires 20%–60% of available roofs larger than 2,000 square meters to be green roofs, and Shanghai with its 2 million square meters of green roofing to meet heat, pollution, and flooding challenges. In Melbourne, Australia, a green factor assessment tool creates new buildings, including green infrastructure. Singapore is promoting regulations that imply site greenery, such as green facades and sky gardens, equivalent to the size of the developed site, and aims to double its sky gardens by 2030. Rooftop gardening, often initiated by the users or inhabitants of the building, is an increasing global movement.

Nature-based solutions are drivers to increase the sustainability and circularity of the construction sector, whereas liveability improves in the built environment, especially in the urban context. Increasing the adoption of nature-based solutions requires the targeted provision of information to the clients and users of the building concerning the benefits. In addition, best practices between clients and construction professionals must be shared. This would significantly increase the sector's share of sustainability, circularity, and liveability.

Nature can serve as a source for the bioinspired design of buildings with energy efficiency, carbon footprint, and passive design in mind. The main design principles of nature-based solutions include bioclimatic, biophilic, and biomimetic approaches (see below). Of course, most adaptations for the built environment are combinations of these three approaches and the average mainstream design.

Bioclimatic approach

The bioclimatic design approach originates from vernacular and indigenous traditions of the design and construction of buildings, as these traditions have created an optimum outcome for locally adapted buildings and settlements. Loftness (2020) defines its content as '...an approach for the design of buildings and landscapes based on the local climate. Bioclimatic design techniques include solar heating and sun shading, natural ventilation, and the use of building materials for thermal time lag and storage'. This approach can be applied to buildings, cities, other urban environments, landscaping, and regional planning. The principles are to take advantage of the microclimate surrounding the building, such as the sun, wind, humidity, precipitation, and surrounding vegetation, by optimising:

- *Conduction* – The direct contact between hot and cool
- *Convection* – The air flow between hot and cool
- *Radiation* – From hot to cool in direct view
- *Evaporation* – The change between a liquid to a gaseous state
- *Thermal storage* – Heat charge and discharge depending on material-specific heat, mass, and conductivity

The prioritisation of these factors depends on the type or function of the construction project, client objectives, site, site microclimate, and macro climate conditions. The key measures for managing these principles are as follows:

- Minimise conductive heat flow by applying insulation in the envelope as a thermal brake to alleviate significant temperature differences between outdoor and indoor climates.
- Manage periodic heat flow by delaying the thermal impact through the selection of material layers.
- inimise thermal bridges between materials in the envelope and air leakage through cracks and around windows and doors.

- Optimise the use of thermal storage by selecting materials with high thermal capacity inside the envelope to equalise the temperature differences between day-time and night time.
- Manage passive solar impact by promoting solar input into thermal storage through windows and greenhouses during winter and minimising the impact of shading equipment and envelope insulation during summer. Different types of glasses in windows can promote either or both the desired functions.
- Minimise the convection impact during winter by integrating barriers for the dominant wind and shaping the building appropriately for wind conditions at the site.
- Manage the indoor climate by using natural draft ventilation, either with cross-flowing air from the wind or by using the differences in stack pressures between outdoor and indoor as a 'chimney effect'. A PV-driven fan could also effectively enhance these methods.
- Manage cooling demand using the evaporative technique by moisturising the air-flow through ventilation or by spraying the envelope. Another technique is to use radiant heat exchange during night time.

Biophilic approach

A biophilic design approach briefly addresses the connection between nature's experience and the awareness of human health and well-being. Some standards and certification systems use biophilic design as a tool to contribute to better indoor environment quality, including human health and well-being issues, such as work-place stress, student performance, patient recovery, and community cohesiveness. Regarding the built environment, Loftness (2020) describes the three major classifications often used as: nature in the space, natural analogies, and nature of the space. Nature in space includes the direct presence of nature in a space or place represented by elements such as plant life, animals, wind, sound, and scents. Specific examples include flowerbeds, bird feeders, water features, fountains, green walls, and vegetated roofs. Natural analogies refer to the features of objects, such as materials, shapes, colours, sequences on decorations, artworks, furniture, and textiles in the built environment. Lastly, the nature of the space relates to how humans feel about natural experiences when in different modes, e.g. relaxing, reducing stress, pleasant, and increased concentration. These classifications were divided into 14 biophilic design patterns as follows:
Nature in space is divided into seven patterns and includes:

- Visual connection with nature
  o Possibility for humans to view natural elements, living systems, and natural processes, such as shade trees, flowering plants, and bodies of (clean) water. Impacts stress, positive emotional functioning, concentration, and recovery rates.
- Non-visual connection with nature
  o Sounds, aromas, and textures reminiscent of outdoor environments using human sensory systems such as auditory, olfactory, haptic, and gustatory sensor systems, e.g. ocean waves, herb scents, pet therapy, and tasting. Impacts blood pressure, cognitive performance, mental health, and tranquillity.

- Non-rhythmic sensory stimuli
  - Stochastic movements of living objects or things immediately engage human sensor systems and, compared to mechanical movements such as a ticking clock, tend to capture an individual's focus for a longer time, e.g. butterfly movement, rustling leaves, chirping birds, and falling water. Impacts eye lens relaxation, heart rate, blood pressure, attention, and exploration behaviour.
- Thermal and airflow variability
  - Subtle changes in air temperature, humidity, surface temperatures, and airflow on the skin mimic natural processes, whereas mainstream indoor climate design aims to minimise the variability of indoor climate. Impacts comfort, well-being, sense of pleasure, attention, and concentration.
- Presence of water
  - Experiences that facilitate seeing, hearing, or touching water through fluidity, sound, lighting, proximity, and accessibility. Impacts stress reduction, tranquillity, mood, heart rate, blood pressure, concentration, and memory.
- Dynamic and diffuse light
  - The dynamic intensity of light and shadows change over time, mimicking natural conditions. Impacts mood, performance, well-being, body temperature, heart rate, alertness, and so on.
- Connection with natural systems
  - Awareness of seasonal and temporal changes in natural processes, especially of a healthy ecosystem; evokes relationships with the whole, e.g. awareness of seasonality and cycles of life, such as the Japanese love of the very short period of cherry blossoms. Impacts as the presence of water above.

Natural analogies consist of three patterns and include:

- Biomorphic forms and patterns
  - Conveys symbols of contoured, patterned, textured, or numerical arrangements originating from nature. Impacts stress, concentration, and cognitive performance.
- Material connection with nature
  - Materials and elements are minimally processed, reflecting the local ecology or geology aimed at creating a sense of place. Impacts cognitive performance, comfort, blood pressure, pulse rate, and so on.
- Complexity and order
  - Symmetries and fractal structures in coherent spatial hierarchy. Impacts cognitive performance, health, and well-being (example in Figure 6.1)

Lastly, nature of the space comprises four patterns and includes:

- Prospect
  - Unimpeded long-distance view for planning and surveillance, such as orientation of buildings, fenestrations, corridors, and workplaces that optimise visual access of indoor–outdoor places. Impacts stress, boredom, fatigue, and irritation as well as vulnerability and comfort.

- Refuge
  - A place where individuals can withdraw from the main flow of activity, where they are protected from behind and above. Impacts stress, blood pressure, heart rate, irritation, fatigue, concentration, attention, and sense of safety.
- Mystery
  - Teasing and compelling individuals to further investigate with a promise of further information from partially obscured views or other sensory parts, inciting the desire to explore further, e.g. 'What is behind the corner'? Impacts pleasure, curiosity, and interest.
- Risk/peril
  - An identifiable threat combined with reliable safety might be dangerous but irresistible, and worth exploring. Impacts attention, curiosity, memory, problem solving, and control functions.

When implementing a biophilic approach in the design stage, it is crucial to consider the client's objectives and intent, particularly concerning the function of the building and the coming user's expectations and demands. More specifically, issues concerning human health and well-being are the most important. Another issue to be addressed is how biophilic design can improve the client's sustainability policy (focusing on the social aspect) and the environmental management system (EMS) and its commitment to continual improvements (see more in Chapter 3, Section 3.2). To enhance the user's health and well-being, it should be based on the population in question and the place and site of the building. Combining the patterns of biophilic design described above could increase the benefits of health and well-being; adding and combining multiple patterns (that are not mutually supportive and integrative) merely for the sake of increasing diversity could be counterproductive. As general guidance, empirical studies show that benefits from, for example, positive emotions and mental restoration can occur within 5–20 minutes.

Biomimicry approach

Biomimicry design is an approach in which a nature-based solution mimics a natural bio-function. An important factor that differentiates biomimicry from other nature-based design approaches is the emphasis on learning and elaborating on regenerative solutions that living systems have for specific functional challenges. The differences between biomimicry, biomorphism, and bio-utilisation are commonly misinterpreted. The latter refers to the use of biomaterials as a technical solution, such as trees to make furniture or a living wall of plants to purify indoor air. When designing a space with elements that visually resemble nature, biomorphism applies a 'look like nature' rather than 'work like nature' approach (biomimicry), see example in Figure 6.2. Conversely, biomimicry design could be far from visually similar to natural elements; thus, functionality is an important factor. Nevertheless, there are many examples of designs in which biomorphism and bio-utilisation elements are used in biomimicry approaches. Biomimicry is characterised by studying and emulating living system functional strategies to create sustainable regenerative solutions and adapt to climate change while also embodying and (re)connecting elements of biology in buildings and in the built environment. Biomimicry design may not be similar to the source of an organism/ecosystem but entails the same functional concepts.

Figure 6.1 Example of complexity and order, Hays Galleria, London, UK (Author original)).

Studying and mimicking nature with the aim of developing practical solutions that address human needs is not a novel practice. In the past, nature inspired humans to provide food and shelter, and to develop methods to survive harsh environments. Such methods were later reapplied in modern society in sectors such as the built environment, medical science, defence, agriculture, and even manufacturing processes. Ancient Greeks used biomimicry to develop the concept of classical beauty from natural organisms through mechanism, shape, and functions. Leonardo da Vinci also used biomimicry to create the concept of the flying machine by studying birds' ways of flying, which was later used to develop the first flying vehicles at the beginning of the 1900s. More examples of biomimicry approaches in design include the Casa Batllo building by A. Gaudi from the late 1800s in Barcelona, Spain, which drew inspiration from a human skeleton; the Eiffel tower in Paris, France, which features the structural function of the human thighbone; and the National Stadium in Beijing, China, which was modelled and structured to resemble a bird's nest.

Figure 6.2 Example of a biomorphic timber structure, Las Setas, Seville, Spain (Author
      original)).

The mimicry utilised by biomimetic design approaches could be a problem- or
solution-driven approach. In the first approach, the designer studies solutions to address
problems through biology, and how organisms and systems in nature have solved similar problems. Conversely, in the solution-driven approach, biology is used as a solution
to copy and then transfer to the design problem. A possible disadvantage of the problem-driven approach is that it determines the correlation between buildings and the ecosystem of which they are a part; thus, the underlying causes of a possible unsustainable or
degenerative built environment are not necessarily addressed. While a problem-driven
approach may be a start to begin transferring a built environment to a more sustainable state, a solution-driven approach involves the influence of biological knowledge on
human design. An advantage is that the knowledge of biology can influence the design
in ways other than the original design problem. A disadvantage is that when in-depth
biological research is required, the result must be relevant to the design context.

Biomimicry can be divided into three levels: organism (imitation of nature), behaviour (imitation of natural processes), and ecosystem (imitation of the working principles of ecosystems). At the organism level, building design is primarily inspired by the

form, shape, or structure of an organism. The designer analyses the form and function of a specific organism, and then chooses to mimic a part of or the entire organism. At the behavioural level, the interaction between the ecosystem and its surroundings inspires the design of the building and its environment; it is when the analysis involves how an organism interacts with its immediate environment to design a structure that fits in the original design problem. At the ecosystem level, the focus is on how different parts of an organism or many organisms interact and transfer to a larger urban scale. This level involves mimicking how an organism interacts with the environment and how several components work together and is at an urban scale or in a larger complex project with multiple elements.

Summary of biomimicry design levels:

- *Organism level* – Mimicry of a specific organism
- *Behaviour level* – Mimicry of the way that an organism behaves or relates to its larger context
- *Ecosystem level* – Mimicry of an ecosystem

Five different dimensions of mimicry offered by each of the three levels:

- How the design mimics the appearance and form of an ecosystem
- How it mimics the material of an ecosystem
- How it mimics the way that the ecosystem is being constructed
- How the ecosystem works, what the process looks like
- What the ecosystem is capable of, the function of the system

Examples of biomimicry design in buildings:

- The Eastgate Centre in Harare, Zimbabwe (see Figure 6.3), was designed based on how termites construct their nests to maintain temperatures independent of exterior weather. The result is a peppered design with holes all over the building's envelope that requires minimal active cooling and ventilation for thermal comfort (behaviour level).
- The Gherkin building in London, UK (see Figure 6.4), was inspired by sea sponges to increase structural strength and drastically reduce energy requirements for ventilation, cooling, and heating. Similar to sponges, open shafts were constructed between each layer to enable passive heat exchange and penetration of sunlight deep into the building (organism level).
- The Bio-Intelligent Quotient House in Hamburg, Germany (see Figure 6.5), is another example of nature-inspired building design: a passive house generates energy using algae biomass harvested from the building's façade, built of water-filled windows. The facade absorbs light that is not used by the algae to generate heat for hot water and heating (behaviour level).

Nature-based solutions and ecosystem services

Ecosystem services comprise the benefits humans receive either directly or indirectly from ecosystems. Humans are entirely dependent on ecosystem services for their

Figure 6.3 The Eastgate Center, Harare, Zimbabwe (Chayaamor-Heil, N. and Vitalis, L. (2021)).

survival, well-being, and economies. However, it is well known that urban environments have a significant negative effect on ecosystems and the benefits they provide.

One way to reduce this effect over time is to regenerate urban areas so that they provide, integrate with, and support ecosystem services. As mentioned in the previous chapter (see Chapter 4, Section 4.2), ecosystem services are divided into provisioning, regulation, supporting, and cultural services. The latter includes tourism, recreation, appreciation, and education.

If a building and its environment provide its own ecosystem services, the pressure on local and distant ecosystems could decrease and be able to regenerate to support more species by increasing biodiversity. Regenerative design is used to address the built environment to restore the capacity of ecosystems functioning at optimal health for the mutual benefit of both humans and non-humans. Instead of viewing a building as a stand-alone object, it is considered as a node in a larger system, similar to a single organism in an ecosystem. From this perspective, mutual and beneficial interactions between a building, its environment, and its human users or inhabitants could occur. An example is when the target of water consumption is determined as relative to understanding the surrounding ecosystem, i.e. using a target based on annual rainfall budget instead of the target of reducing average consumption by 10%. The latter figure is based on the targets of economic, political, or convenience factors and is not related to a sustainable system or whether it is physically possible for the actual site. A common solution to such problems regarding consumption is to decrease, remove, or stop human behaviours or ways of designing the built environment. On the contrary, the first figure allows the possibility of using or taking advantage of a renewable source for the actual site physically to make it in measurable units and, further, to become

Figure 6.4 The Gherkin building, London, UK (Pixabay (2022)).

understandable in an ecosystem context. This kind of regenerative approach enables a healthier interrelationship and more healthy building site rather than only reducing the negative impact on the environment.

Designing buildings and their surroundings with an approach that mimics nature and the behaviour of an ecosystem provides ecosystem services. Knowledge of site-specific targets from an ecosystem perspective enables potential integration with and contributions to existing ecosystems, rather than depleting them. However, if a

(a)   (b)

Figure 6.5 The BIQ-House Hamburg, Germany (Biloria, N. and Thakkar, Y. (2020)).

building or its specific site take advantage of the provided ecosystem services with-out contributing back to ecosystems or producing them, the building and its site will degrade the climate and ecosystems. Thus, it is very important to reverse this process and to contribute to ecosystem services through the built environment by implement-ing practices such as:

• Developing systems that provide habitats for species adaptable to live and coexist with humans in the built environment
• Developing systems for soil and fertility renewals by carefully selecting biode-gradable waste and non-biodegradable waste for recycling purposes
• Purifying air and water contaminated by the urban environment
• Regulating climate impact by reducing carbon emissions and using the heat island effect
• Equestering carbon by carbon-sink technology
• Producing local renewable energy
• Collecting and distributing fresh water through local rainwater collection systems
• Locally producing human food through urban ecological agriculture

Generally, viewing the system as a whole, all aspects of ecosystem services are important. However, with regard to the numerous (often more than 26 different) identified ecosystem services and the limitations of actual built environment areas, an ecosystem services analysis (ESA) must be performed (see more about ESA in Chapter 4). Per Pedersen (2017), there are approximately seven different ecosystem services applicable to the urban environment: habitat provision, nutrient cycling, purification, climate regulation, provision of energy for human consumption, pro-vision of fresh water, and provision of food. These ecosystem services applicable to the urban context do not include the cultural ecosystem services mentioned in Chapter 4. However, it is essential to consider humans and their relationship with nature in terms of socialisation, stress management, handling emotional states, managing anger and violence, promoting recovery from illness or trauma, and increasing the rate of concentration and productivity. Measuring and analysing these kinds of cultural ecosystem services is quite different from dealing with provisioning, supporting, or regulating ecosystem services, and is, therefore, not included in an ESA.

Timeline for new buildings with zero carbon emissions and sustainability

The actions presented in this subsection offer a path to increase the performance of new buildings to become zero emission, efficient, and resilient as soon as possible by utilising key tools of beneficial policies at national and subnational levels. Concerning a roadmap for existing buildings, see the section on operation and maintenance. Below, it is a list of some of the target actions to be met. The baseline for the target setting, expressed as current, derives from the actual status of 2020–2022. The expression 'short term' refers to the ten years that follow, with the deadline of 2030, following 'medium term' with the deadline of 2040, and, finally, 'long term' with the deadline 2050. (This roadmap was established by the Global ABC, IEA, and UNEP 2020; see the list of references at the end of this chapter.) The target actions are as follows:

*   *Energy codes* – Minimum requirements regarding building performance or building components to achieve zero carbon emission, efficient, and resilient buildings. This is also a requirement for upgrading major renovations of existing buildings. The codes should be based on whole lifecycle carbon emissions, including embodied and operational carbon emissions. The priority of coding should be the highest efficiency to the lowest cost and locally adapted bioclimatic design principles to optimise passive design. To begin, codes should be in a prescriptive format and later evolve into performance-based.
    o   Currently, only one-third of world countries have mandatory codes
    o   The timeline of the target should comprise short-term mandatory codes for most countries, with some volunteers with near-zero targets
    o   The medium-term target should strive for most countries to have near zero codes
    o   The long-term target should strive for all countries to have near zero carbon emission codes
*   *Compliance of the codes* – Compliance and enforcement assurance of the building codes is crucial but challenging for the assessment bodies, especially in the informal building sector. A framework to monitor compliance checks, accessible guidelines and tools, and extensive capacity building ambition, together with sufficient stakeholder engagement, should facilitate levels of compliance, especially enabling compliance in the informal sector and in social housing.
    o   Currently, there is a lack of enforcement and monitoring of compliance
    o   In the short term, a monitoring framework is in place, tools to enable simplified compliance are available, and more than half of the buildings are compliant. Most countries monitor their informal sectors.
    o   The medium-term target requires most new buildings to be compliant with codes, and most of the informal sector is compliant with codes

- o The long-term target requires that nearly all buildings almost everywhere are compliant with codes and meet minimum standards of requirements
- *Building labels and certificates* – Building energy labels can be used to confirm a building's performance as a quantitative measurement, scaled from poor to good efficiency. This also enables stakeholders' awareness of the performance of the building and the possibility of connecting to financial tools and incentives. A certificate from a 'green' building certification strengthens the assessed building's performance. The labelling and certification systems should be continually monitored and improved; the highest score should always be the frontrunner.
  - o Currently, only a few buildings receive voluntary labels or certifications
  - o In the short term, it should be mandatory that half of the new buildings have labels
  - o The medium-term target should require it to be mandatory for most new buildings to have labels
  - o The long-term target should make it mandatory for all buildings to have a label
- *Labels of building components* – Access to credible and reliable information concerning the content of different building components and materials is key to designing buildings to meet sustainability, resilience, and zero carbon performance. It should include parameters such as the thermal transmittance of materials, solar factor or heat gain coefficient of glazing, embodied carbon content (especially for materials with significant levels of embodied carbon), and hazardous and dangerous chemical content in materials that affect human health and biodiversity.
  - o Currently, minimal information is available on the performance of building materials and components
  - o In the short term, it should be mandatory to label the main components, including the carbon content
  - o The medium-term target should require that all countries comply with mandatory labelling of the main components, including carbon content
  - o The long-term target should require mandatory labelling concerning embodied carbon and components performance
- *Building passport* – A document or logbook used to track information about systems, energy performance, improvements by maintenance or renovations, and components used for a certain building (see Chapter 1, Section 1.6). The use of passports improves a client's decision-making processes during the building's life cycle. During the transfer of project documents between the contractor and client at the end of the construction project process, a basic building passport should be established and could include the floor area, material logbook with specifications, quantities, content of hazardous components, and content of embodied carbon. Furthermore, it could include the description of the systems used, supposed energy performance, and estimated maintenance schedules. During the operation of the building, passports may be completed with further and actual information.
  - o The current global status is the limited use of building passports and minimal collected information
  - o From a short-term point of view, passports should be widespread, with basic information, including the content of materials and embodied carbon
  - o The medium-term target should require that half of the new buildings have a full passport and that all buildings convey their embodied carbon content

o In the long term, building passports are widely used, including full data on all used materials
- *Lifecycle analysis (LCA)* – When deciding on building usage, concerns may include the whole life of the building, from the early conceptual and briefing stages through design, construction, and operation, including maintenance and refurbishing until the end of life, deconstruction, and disassembling. National or regional databases containing information on embodied carbon in construction materials are essential when choosing design solutions with proper lifecycle impact analysis.
  o Currently, LCA is minimally available and used
  o For the short-term period, LCA is mandatory for most new buildings and the availability of LCA data for the main construction materials appears in national or regional databases
  o The medium-term target should strive for complete databases for all construction materials and mandatory LCA for all new buildings
  o In the long term, there are comprehensive databases for all construction materials that support mandatory LCA for all buildings

Various fiscal incentives to promote the above roadmap could include awards to the best-performing buildings for using the most effective new technologies and tools. A non-fiscal incentive to promote construction projects to encourage high-performance buildings could be expedited building permits or increased allowances for floor areas.

Key factors of the building envelope

The performance of the building envelope, including structure, insulation, conduction, and radiation heat transfer through outer walls and windows, can be measured by the overall thermal transfer value (OTTV). A lower OTTV can be achieved through optimised material choices and passive design strategies, such as building form, orientation, increased thermal mass, shading, the use of reflective surfaces to limit solar gain, and the use of vegetation, e.g. in green roofs. Passive design strategies, such as a cost-effective combination of the thermal performance of the building envelope, solar gains, ventilation system, and indoor impact of solar radiation, are highly dependent on the building type and how it will be used. Passive design alternatives must be developed for each building, depending on the bioclimatic conditions of each building site. Concerning the OTTV, insulation is one of the components that should have specific targets depending on the building's geographical and regional location, and hot to cold locations. The thermal performance of a material depends on its thermal conductivity, which is expressed by the U-value, that is, the amount of heat transferred through a given thickness of a specific material. The lower the U-value, the better the material acts as an insulator. Good insulation capacity is important in both hot and cold climates. It is most effective as a component of the most expansive surface area, such as the roof of low-flat buildings or external walls of tall buildings. The insulation benefit as an energy saver must be calculated and optimised with respect to the embodied carbon content during the entire life cycle of the insulation material.

Heat transfer by conduction through windows can be reduced through the use of double- or triple-glazed windows, which have lower U-values. Such windows also provide noise protection, improve indoor thermal comfort, and facilitate passive design strategies. Solar radiation, the dominant source of heat transfer through the windows, can be reduced using windows with low emissivity and a low solar heat gain coefficient. Building design and new technologies could simultaneously enable low solar heat gain and allow natural daylight during hot weather. Access to views and daylight is also essential for the well-being, health, and productivity of building occupants. A good design should ensure that all spaces in a building have access to natural light, favourable views, and optimum daylight levels for most of the day. However, it is also crucial to balance and optimise the occurrence of natural light by controlling severe solar radiation. Another issue of balancing incoming solar radiation is to ensure the use of reflective surfaces, such as a light colour (or colours) with reflective pigments, on the surfaces most exposed to direct sunlight (mostly the roof). However, the most cost-effective method is to reduce the impact of solar radiation by decreasing the window size and providing shading. Good shading can have the same effect as solar performance glazing in reducing solar radiation impact. External shading in the form of horizontal, vertical, fixed, or movable elements is probably the most cost-effective method for blocking solar radiation. Of course, all these beneficial measures must be assessed with respect to embodied carbon from a whole lifecycle perspective.

Key factors of HVAC and appliances

Energy-consuming appliances and equipment systems, particularly for heating, cooling, and ventilation, have a shorter lifetime than the building itself, offering a significant opportunity to reduce carbon emissions. In addition to increasing the efficiency of the appliances, factors of human behaviour must also be considered, such as how the users operate the appliances, e.g. setting air conditioners at temperature set points that are lower than required.

EXTERNAL

Controllable and efficient ventilation with energy retention is essential for improving indoor air quality. When ventilation efficiency is increased using mechanical, natural, and hybrid ventilation types, buildings can shift to hybrid ventilation, which uses natural ventilation when feasible and mechanical ventilation when natural ventilation is not effective. With further improvements, the system should include energy recovery to enable air exchange with minimal heat and humidity transfer. The system energy recovery efficiency should also require that systems be improved from low efficiency (about 50% efficiency) to high efficiency (in the 80%–90% range).

Space heating systems can enable a more efficient delivery of indoor comfort through improved system efficiency. These heating systems could also offer opportunities for low-carbon or zero carbon emissions during the transition from fossil fuel heating systems to electricity or renewable energy heating systems. The key technologies to achieve these reductions include heat pumps, modern biomass stoves and boilers, phasing out traditional biomass, and the use of waste heat or co-generation. Regarding water-heating systems, the use of modern renewable energy sources and improved system efficiency can enable efficient hot-water delivery. Heat pumps, solar thermal water heaters, modern biomass boilers, and the use of waste heat and co-generation offer solutions for low- or zero carbon-emission water heating.

Thermal energy storage for heating or cooling enables load shifting and optimised heat transfer efficiency. This could be important with regard to the increasing pressure on peak electricity demand due to growing electrification. Energy storage could be achieved by highly insulated water or refrigerant tanks, thermal mass, or phase-change materials of the construction. Electrical storage by batteries could also become important in terms of decentralised renewable electricity generation and the demand for interconnectivity between electric vehicles and buildings.

Concerning appliances, there is a need for more efficient large and small appliances to counter the vast increase in usage depending on the growing affluence. One example is a refrigerator with improved compressors with variable speed, insulation, and heat-pump efficiency. Smart or connecting user controlling devices by demand-side response could manage appliance demand efficiently, reducing standby losses and connectivity by sensors, enabling low-power mode, load balancing, and remote programming.

COOL BUILDINGS

Space cooling systems are the fastest growing end-use systems in buildings. Approximately, 2.8 billion people live where average daily temperatures exceed 25° C all year, and only about 8% of them are using mechanical air conditioning systems (AC). Driven by rising incomes, an expected increase of 75% in building floor area by 2050 (80% in developing economies), a warming planet with higher temperatures, more frequent heat waves, and rapid urbanisation in the coming 30 years, the need for space cooling could triple, especially in hot and tropical regions. Residential buildings are responsible for over two-thirds of this increase. A vast amount of building space around the world is being constructed with a small capacity to adapt to the surrounding climate. Constructed from steel, concrete, and glass, buildings can overheat without sufficient thermal breaks, shading, ventilation, or insulation. However, to provide indoor thermal comfort, it is necessary to use excessive energy for mechanical cooling. The energy used for space cooling is expected to double current levels by 2040. According to figures, space cooling is responsible for significant energy use and emissions, contributing to approximately one gigaton of carbon emissions and nearly 5% of total energy consumption worldwide in 2020. These circumstances provide an opportunity to reduce the demand of cooling energy by applying bioclimatic design and passive building design principles and to emphasise climate-adapted building designs, construction, and operations aligned to the concept of 'Avoid-Shift-Improve' (see below). For a net zero energy scenario, energy use for space cooling must be reduced 50% by 2050.

Furthermore, with advanced cooling technology with an improved peak demand energy efficiency ratio and seasonal energy efficiency ratio, the systems can deliver high thermal comfort. Alongside design strategies that minimise the need for cooling, an adoption of hybrid cooling methods, such as evaporative cooling, ventilate cooling, and other 'free cooling' that uses ground or water temperatures, can support the increased overall efficiency. Overall system efficiency also increases with the use of variable-speed drives and improved thermal distribution efficiency.

Designing cool buildings can be a solution where it is possible to reduce or even avoid the demand for cooling appliances. The use of climate adapted envelopes, including specific exterior colours, windows, natural ventilation, orientation, and vegetation, can reduce the energy demand of mechanical cooling. However, traditional buildings in hot climates often achieve comfortable conditions without mechanical cooling, for example, long roof overhangs, exterior shading elements, and green court-yards provide shade to buildings, reducing the impact of solar heat. Nevertheless, the use of white roofs or façade coatings can reflect 80% of the solar radiation compared to those that are dark or black. Furthermore, by using high-performance insulation in the building envelope, the possibility of energy reduction from cooling demand can extend up to 30%–50%. The use of windows with low-emissivity glass that reflects infra-red solar radiation without affecting incoming daylight can reduce it further by approximately 20%. Thus, windows should be shaded with external shading devices and the window-to-wall ratio should be adapted to the climate zone to ensure suffi-cient natural daylight. From analysis of Thailand and China, a natural ventilation approach can reduce the cooling demand from approximately 8%–40%. Finally, well-designed residential landscapes and vegetation could reduce the cooling demand by approximately 25%.

When buildings are adapted to the local climate and cool building approaches are implemented, mechanical cooling and natural cooling can be avoided. There are many variations depending on the site of the building, the local climate, local build-ing culture, and use of the building. However, two principles can be applied in humid climates – using light - to mid-weight structures and open spacious designs that allow continuous natural ventilation. In dry climates, massive structures can be used to allow protection from heat during the day and, if nights are cool, conserve the heat, or naturally cool it by night.

By following the steps 'Avoid- Shift- Improve' (see below) to design cool and low-carbon buildings adapted to the local climate can reduce or even avoid the need for cooling and increase indoor comfort. However, this can be applied only from the beginning of the design process.

Avoid – High cooling demand through building design adapted to local climate:

- By adapting nature-based solutions and passive building design
- Through site adaption
  - By taking advantage of the site's surrounding vegetation, water bodies, and neighbouring buildings for shading and cooling
  - By reducing urban heat islands close to the site with measures such as green roofs, broad-leaf trees, and bushes providing shading without disturbing air-flow around the site
- Through orientation and shape
  - Orientation of the building from east to west, avoiding exposure to facades from solar radiation with low angles
  - Employing the main principles of building design in humid and dry climates, as mentioned above
  - By facing the doors northward and optimising window placement by daylight and using low-emissivity glass
  - By optimising the window-wall ratio, in hot climates not exceeding 20%
- Through the building envelope
  - Massive walls in a hot climate to keep heat out during the day and to slow release the absorbed warmth at night, or light walls in a humid climate with many openings to promote natural ventilation
  - Insulation of the roofs, massive roofs in hot conditions, and light roofs in humid conditions

o Implement shading facades with roof overhangs and exterior shadings
o Bright and reflective coatings and/or green vegetation on roofs and facades

Shift – Remaining cooling demand with renewables, thermal storage, and district cooling

- By using energy from renewable sources, commercial sources from the grid, or produced onsite by PVs
- By using solar-powered cold chains or stations, an off-grid solution for the storage and preservation of delicate products in remote areas
- By using district cooling facilities where available

Improve – Conventional cooling with highly efficient and low-carbon appliances

- By using ceiling fans primarily instead of mechanical devices – because of less energy and a significant increase in thermal comfort, it is equivalent to a reduction in indoor temperature by two degrees
- By using digital technology to control and optimise the cooling distribution and demand
- By avoiding appliances with harmful refrigerants such as CFCs and HCFCs. By using alternative refrigerants with low or no greenhouse impact, there is a greater decrease in direct emissions, but due to this transition, considerations must be made regarding the thermodynamic performance in cooling as well as flammability and toxicity.

As mentioned earlier in this section, there is an immense need for improved building designs with efficient nature-based solutions and for cooling technologies and appliances that consider the climatic and cultural context. A step towards this is to follow the advice of the 'Avoid-Shift-Improve' steps.

Key factors of healthy homes in the built environment

Ensuring access to sustainable and healthy homes in the built environment also improves public health, reduces inequalities and carbon emissions, and improves urban sustainability and resilience. Ongoing urbanisation, demographic changes, world population increase, and climate change create an immense demand for housing which meets the global demand for future zero carbon buildings. Challenges such as 3 billion people requiring access to adequate housing by 2030 or a doubling of the world population aged 60+ by 2050 creates an immense global demand for new housing, as does an increase in extreme weather events due to climate change.

Housing – where people live, sleep, and work – is a central element of society. People working in their homes has become increasingly important as a result of the recent Covid-19 situation, from 2020 to 2021 and in the foreseeable future. Human health is connected to housing as a health determinant. Healthy housing saves lives, prevents diseases, and enhances human well-being and is a part of the UN Sustainable Development Goals (SDG), including Health (SDG 3), Reducing Inequalities (SDG 10), Achieving Clean Energy (SDG 7), Sustainable Cities (SDG 11), and Climate Action (SDG13).

Per the World Health Organization (WHO), the structure of a residence, social environment, neighbourhood, and community are four dimensions of healthy housing.

Poor housing conditions increase the risk of both contagious and non-contagious diseases, such as difficulties with or the high cost of heating a residence, which can cause respiratory or cardiovascular diseases. However, high indoor temperatures can also cause cardiovascular diseases. Residences with a very high number of inhabitants, which reduces the available space, can increase the risk of contagious diseases. This was especially relevant during the recent Covid-19 pandemic when sufficient space was needed between individuals to reduce transmission and increased noncircular air ventilation decreased the risk of the indoor spreading of the virus.

In general, poor indoor quality can also be associated with pneumonia, stroke, and lung cancer caused by indoor sources such as air pollution from solid fuels (coal, wood, or dung) using insufficient open stoves, or by exposure to naturally radioactive radon gas. Access to safe water and sanitation services is also crucial for healthy housing; billions of people still lack handwashing facilities with soap and water in their homes, as well as access to basic drinking water. Poor accessibility, especially for the elderly and people with disabilities, could be a risk factor for physical injury, stress, and isolation as well as for places with structural deficiencies.

Access to daylight is essential for individual well-being, health, and productivity. The design should ensure that all spaces have access to natural light and views and provide glare-free, adequate daylight levels during the daytime. During other times of the day, delivery of visual comfort must be secured through energy-efficient lumens per watt with improved lighting technology, such as intelligent controls, sensors, and shading devices.

Socio-economic factors, such as low income, neighbourhood conditions, and social inequality, together with environmental factors, such as indoor air quality, toxic building materials, and structural insufficiency, are basic components of housing inequality. Furthermore, tenure insecurity can affect physical and mental health. These factors indicate whether residents are able to afford and maintain safe and healthy housing. The cost of maintenance can be related to the cost of safe drinking water, electricity, fuel for heating, cooking, and lighting. It is estimated that nearly 2 billion people live in unhealthy homes, including slums and other informal and overcrowded settlements.

The WHO has developed the WHO Housing and Health Guidelines (WHO, 2018) with recommendations on how to address unhealthy residences, such as living space (crowding), low and high indoor temperatures, injury hazards in the home, and accessibility of housing for people with functional disabilities. It also contains a summary of existing WHO recommendations regarding housing, water, air quality, noise levels and frequency of asbestos, lead, tobacco smoke, and radon. The implementation of the guidelines could be at the national, sub-national, or local level, if the former will, according to WHO, 'make a substantive contribution to the provision of healthy and sustainable housing for all'.

Many of these issues regarding indoor health, visual, acoustic, and thermal comfort and occupant well-being are not all suitable for objective measurement metrics; on the contrary, they require a set of subjective metrics, mostly from post-occupancy surveys, POEs (see more about POE later in this chapter, in Section 6.3). Generally, six main variables significantly impact user satisfaction (listed below, with the suggested measurement metrics):

- Occupant density, by square meter per person appropriate to building type
- Comfort, including appropriate metrics

- o personal controls, by time of response
- o thermal comfort, by temperature
- o indoor air quality, by CO_2, CO, NO_x, TVOCs, VOCs, or concentration of mould spores
- o visual comfort, by average daylight factor
- o acoustic comfort, by noise rating and reverberation time
- o contact with nature by opening a window within 7 m with biophilia (see above about biophilia) such as contact with views, places, plants, and natural materials
- Responsiveness to need, including comfort (from occupant density), but a host of other ways in which needs should be met effectively, usually from post-occupancy surveys
- Ventilation type, which also encompasses attributes such as size, building depth, and other allometric properties, i.e. how size affects shape, volume, and services
- Workgroups and their layout in the space plan
- Design intent and how this is communicated to users and occupants

It is very important to consider indoor health issues and sustainability holistically during the design stage; for example, achieving net zero carbon emissions must not be contradictory or incompatible with occupant health matters. Various key design principles could be to provide responsive local controls, e.g. opening windows or local control of HVAC systems, and designing spaces with:

- Strong visual connection to the outside
- Appropriate occupant density for activity
- Good indoor air quality
- Good indoor daylighting, lighting, and glare
- Good thermal comfort
- Good acoustic comfort
- Inclusive and universal accessibility
- Active circulation routes, e.g. stairs, bicycling, and walking routes
- Plants, indoor and outdoor

Regenerative design approach

The idea of considering the whole surrounding built environment when designing a construction project – a regenerative design approach – instead of looking at the actual single building and its inherent components as a holistic approach, is a concept that promotes looking beyond sustainability and beyond zero carbon emissions. A regenerative built environment is based on the uniqueness of a place, community engagement, and the continuous creation of a story of the place within the community. When concentrating only on building energy consumption or carbon emissions, it could risk shifting the environmental impact of the building from one factor to another. In contrast, regenerative design principles consider the built environment as a whole with no sub-optimisations.

Regenerative design aims to achieve net positive development with a whole-systems approach that can generate synergies between ecological, economic, and social dimensions. The design approach considers the comprehension between the construction project and its site, the place, throughout the building's entire life. It requires a whole living systems approach, interconnectedness with nature, focus on the location, and the participation of all stakeholders, and enables continuous replacement, renewal, and rebirth. It is how the whole construction process – building design, construction, and operation – positively influence the social, ecological, and economic health of the place within which the building exists, using the health of ecological systems as a basis for design that supports the regeneration of existing and lost ecosystems. The regenerative design approach creates a development that could restore health to both human communities and the ecosystems of which they are a part and engages all stakeholders and processes of the place based on the co-evolution principle, that is, continuously learning how humans can participate in the surrounding environment. This design approach is more complex than ordinary construction design; although it covers all green and sustainability criteria, the difference is that it replaces linear processes with cyclical and requires a shift from prescriptive to descriptive metrics, a shift away from rules and regulations, from product- to process-based, and from linear to non-linear thinking. It requires a more qualitatively different framing of design that complements and supplies the currently available mainstream green building certification tools.

Once the desired patterns of the relationship between the construction project and its place, keystone species, and key system are understood in general terms, the metrics and ways to measure improvements can be established. One way to achieve this is to use the method of ESA mentioned above to identify indicators. However, this identification and assumed understanding is not a completely ensured truth of the whole, or that the involved stakeholders engaged in the project will interact in the assumed way. Therefore, it is essential to continuously monitor the work to achieve the necessary feedback to allow the evolution of the system. This feedback process supports conscious engagement and deep relationships between people and places over time. The process of adapting all parts in relation to the whole requires a few iterations of thinking. However, the method of designing a construction project is a linear process, which depends on a temporal schedule, and it is necessary to make approximations simultaneously with the whole through rapid and frequent iteration of ideas, as a process of integrative design.

The process of regenerative design is iterative and requires improved involvement of the project stakeholders during the design process, providing site-specific information and stories. Further, stakeholders assess the construction project throughout its operation period and keep it regenerative.

The regenerative design process extends throughout the entire life cycle of the construction project, including the operation and de-construction phases. The design work and materials used need to be IT intensive, e.g. using BIM technology. The design needs to be from a 'glocal' point of view, reaching global targets with local adaptation and considerations. The design should be based on story- and place-specific parameters, intensive collaborations with the community, and feedback. The design team must use nature-based solutions as a guide to solve design problems and must be interdisciplinary, consisting of members with different professional backgrounds (e.g. architects, civil engineers, mechanical engineers, computer engineers, electric engineers, biologists, social anthropologists). The design team is responsible for not only an environmentally sufficient design but also the regeneration of nature in the surroundings of the site. The team should have good coordination skills as they must involve both the stakeholders and the community in the design process and knowledge of the IT-based technologies and materials used in the project supporting the regeneration performance of the construction project. For a regenerative design process, it is necessary to establish an interdisciplinary design team in adapting the contemporary technologies into the process.

Three design strategies

One regenerative design approach described by Attia (2018) is based on three design strategies:

- Selection of a construction system
- Defining design elements and their performance
- Choice of regenerative materials

These three basic strategies should be applied at the start of the early conceptual phase of the design process and should be used throughout the entire process. Application of these strategies introduces a new design thinking paradigm, in which sustainability is extensively integrated throughout the entire design process.

When applying the first strategy, the selection and sizing of the construction system are based on the concept of modularity and the possibility of assembling and disassembling different regenerative materials and products. From a sustainability perspective, particular attention should be given to the system to be designed from a disassembling point of view to facilitate adaptability and structural flexibility, adding to an extended presumed lifetime of the facility (see Chapter 2 regarding circularity issues of regenerative design). Based on the construction system concept, the entire envelope of a building can be developed with regard to flexibility and circularity, as mentioned above. Furthermore, the envelope must meet the insulation and hygrothermal requirements of onsite performance conditions using renewable and net-positive impact materials.

Once the construction system has been chosen, the secondary strategy is to analyse the spaces of the building to evaluate nature-based solutions and connection to ecosystem services mentioned earlier. Depending on the microclimate condition of the building site and its geographical location, certain design elements could be more appropriate as regenerative quality markers (e.g. atriums, courtyards, terraces, balconies, skylights, glazed facades, staircases, meeting rooms, open office spaces, common areas, foyers, and roof gardens). The integration of these elements provides quality and a positive impact on end users. The purpose of using regenerative design elements is to improve indoor and outdoor air quality and water usage, and to increase biodiversity, health and well-being of end users, enable cultural and social cohesion, and generate energy. Defining these design elements depends on the type of construction system and scope of the building performance targets and indicators established by the client. Various key areas of interest – such as air quality, human health, energy savings combined with local renewable energy production, water management, and natural design – must be considered when designing a construction project.

When targeting a regenerative positive impact, it is important to improve the indoor and outdoor quality with regard to the operation of the building and the well-being of users. The latter is enhanced by carefully designed availability of natural light and ventilated spaces for living and working, such as gardens, meeting rooms, and common spaces, including staircases. Especially in the indoor environment, it is important to eliminate fine particles in the air, carbon emissions and to regenerate oxygen by using vegetation filters to pass the air through green spaces with a purification intent. The use of different plants and green solutions increases biodiversity and the subjective aspects of the user experience, such as beauty, calmness, and serenity. In urban outdoor areas, air may be purified using green roofs, suspended gardens, vegetated walls, or existing (preferably) or newly planted trees.

Furthermore, it is important to minimise the use of energy and to balance this by producing locally new energy, i.e. by optimising the energy balance to a positive result with no carbon emission impact. To lower energy consumption is to meet the requirements of low-energy building standards, such as Passive House or high rating of green building certifications, e.g. BREEAM, LEED, and so on (more about these and other various green building certifications in Chapter 7). When employing such low or ultra-low energy requirements – less than 15 kWh/m^2 annually – it is essential to design the insulated envelope as airtight as possible, guarantee a minimum fresh air renewal, and avoid overheating during the hot and warm seasons. The envelope outer walls tend to be quite thick due to the requirement of insulation, i.e. the sizing of facades and fenestration must be carefully considered to be suitable for the rest of the regenerative design. Passive solar gains, which promote energy savings, should be optimised on the south façades together with flexible shading devices to prevent overheating during the warm and hot seasons. Concerning the sizing of the window/wall ratio in all orientations, local rules of thumb should be used. It is also favourable to use passive cooling or heating by bypassing and exchanging stable temperature conditions in the soil surrounding the building. When dimensioning a locally renewable energy system, estimations of the amount and size needed that exceed the estimations of building usage of energy should be determined during the early design phase. When choosing the type of renewable system, e.g. thermal, geothermal, photovoltaic, or other systems, the sizing and spatial integration of the building system, form, and envelope must be properly considered. The area intended for PV devices, orientation, and positioning must be studied and estimated. In addition to this study and estimation, the integration and sizing of the facade, roofs, or HVAC systems should be based on and adapted to the location of the building and the building site. The solar thermal system should also be adapted to the need for a hot water supply and stored in insulated tanks to meet the user's needs. The total result of the balanced energy system should be a positive regenerative impact outcome and ensure that the energy surplus produced locally will be integrated into the grid of public use.

Regarding regenerative water usage and management, it is important to collect and harvest rainwater from different wastewater streams. From the perspective of regenerative sewage management, it is optimal to use in situ technology using plant-based purification or phytofilter systems which can treat both grey and blackwater. Phytofilter are plant-based filter systems that consist of plants, special natural substrates, and flowing devices. The filter can have vertical or horizontal flows in both directions and can be used depending on the type of plants and substrates, such as air purifiers or for water treatment, which must be resistant to drought or freezing conditions. Importantly, the users should not use toxic substances to keep the plants alive. This solution also contributes to the greening of the landscape and increasing biodiversity. A very important indicator of regenerative construction is a high and improved supply of water quality, that is, optimised water treatment and nutrient extraction from wastewater. The use of water tanks to collect rainwater during the wet period facilitates the independent usage of water during the warm and dry seasons. Considering size and installation, it should be managed for every individual project, and special attention must be paid to supposed flooding occurrences, especially those predicted and caused by future impacts of climate change.

The concept of design with nature introduces different configurations for vegetation, both indoors and outdoors. These solutions improve the quality of spaces and environmental quality – e.g. biophilia, biomimicry (mentioned earlier), regulation of humidity, and acoustics – and improve external quality, e.g. biodiversity and the heat island effect. Designing with nature starts by connecting green infrastructure with the building and its users to the ecosystem. The well-being of humans is based on a genetic connection to nature, and the use of biophilia approaches should verify this connection. It should be based on balanced nature-based solutions that connect flora and fauna to humans and promote variation in ecosystem services. This could be biodiversity, water management, urban food production, air purification, and human well-being, and also contribute to the recovery of the built environment from heat island effects, acoustic disadvantages, air pollution, and degradation of quality of life. Nature-based solutions include urban agriculture, green roofs, spaces, facades, trees, gardens, parks, ecological networks, and permaculture. Integrating these subjects into the design of a construction project requires careful design and technical studies during the design phase, especially in the early stages. Examples of special assumptions and considerations include root damage, artificial irrigation, structural overload, flow and overflow of water, erosion, light penetration, solar orientation, consequences on envelope insulation, and damp security and plant diversity in order of presumed function and quality. Each construction project must address and integrate these issues individually owing to design objectives, location, site, and microclimate circumstances.

As a summary of this second strategy of regenerative design approach, various examples of important regenerative elements are listed below. It is not an exhaustive list, but rather a list of relevant and commonly used elements to reach a higher level of sustainability towards a regenerative construction project result.

- Windows
- Roof gardens
- Solar panels
- Local wind turbines
- Heat pump technologies
- Solar chimneys
- Greenhouses
- Ventilation chimneys
- Water storage
- Geothermal
- Storage space
- Parking and loading facilities for electrical vehicles
- Phytofilter
- Green walls
- Green facades
- Bio-based insulation
- Green roofs
- Trees, existing and newly planted

The third strategy for designing a regenerative building involves the selection of building materials for the preferred design. This choice must be made without the loss of biological or technical quality. Basic sources should comprise material declared in an environmental product declaration, EPD (more about EPDs, see Chapter 7, in Section 7.1), or as a third-party declaration of materials such as Cradle to Cradle (more about third-party material declarations, see also Chapter 7). These sources of building material declarations should be used in conjunction with the previously mentioned regenerative design principles. However, special attention must be paid to other common technical requirements such as fire safety considerations, structural mechanics, and hygrothermal and acoustic performance. A new focus is to minimise the embodied energy and carbon content. Materials from the technosphere, such as concrete, aluminium, or steel, may be used if they fulfil the requirements as far as possible of a disassembly and reusable design and with a material certificate or declaration as above. Of course, the toxic content in the material must already be excluded as well as that within the production cycles of the material or parts of the material. The preferred material is renewable and natural, such as wood, clay, straw, bamboo, or hemp. Special building construction elements such as foundations, windows, specific safety devices, and technosphere materials are unavoidable and must be declared as above. Specific key questions must be answered during the material selection process of a regenerative construction project's building products, components, or materials:

- Does it fit the purpose?
- Is it available and to what extent?
- Is the raw material available locally?
- How much energy is required to produce the material?
- How much content of embodied carbon is in the material?
- Is material produced locally?
- How much and is there recycling of the residuals from the production?
- Is there an environmental impact or impediment during the production?
- What periodic maintenance is necessary and how is it maintained?
- Is the product ready for disassembling procedures?
- Is the product recyclable?

For a more detailed examination of building materials and the ability to reuse at different levels of design for deconstruction, see Chapter 2.

Twelve design principles

When designing a regenerative building environment, defined criteria must be fulfilled and verified. To date, no clearly explicit criteria have been established for the regenerative design of a construction project; however, Haselsteiner et al. (2021) formulated 12 regenerative design principles derived from various basic regenerative theory literature, concepts from more or less regenerative certifying systems (see Chapter 7), and UN SDGs (see Chapter 1). The 12 principles of regenerative design follow, with various examples of assumed applications:

- *Place, nature, and ecosystem* – Place-based design approach for regenerating the ecosystem and making it possible for future development, such as ecosystem approaches; because each place or site is a unique dynamic entity, the building interacting with site environment and green neighbourhoods, ecosystem and bio-diversity protection, re-establish lost or removed soil, and adapt to microclimate circumstances.

- *Energy* – Restorative and regenerative energy systems, effectively used and shared energy systems, and energy as part of a continuous restoration of ecosystems aiming to increase its quality, such as effectively used and shared renewable, restorative, and regenerative energy systems, and effective energy storage.
- *Carbon* – Carbon-neutral and climate-positive approaches, measures of reducing carbon, such as reducing a building's embodied carbon based on third-party life-cycle assessment sources.
- *Water* – Clean drinking water, such as improved water management and supply and net positive water frequency supply by rainwater harvesting.
- *Material and resources* – Improving material and resource productivity through lifecycle assessment, material from renewable organic systems to conserve resources and maintain them for future generations, such as healthy and non-toxic materials, transparent labelling, the use of local renewables, responsible use, and conservation.
- *Waste* – Zero waste approach, design for disassembly, deconstruction, and flexibility of use, reuse, upcycling and recycling of materials and buildings (see Chapter 2)
- *Health and well-being* – Enhancing the quality of life, reconnecting humans with nature, contributing to individual, community, and societal health and well-being without taking advantage of other people, the environment, or future generations, such as applying the principles of biophilia, ensuring access to healthy food, ensuring air quality, indoor air quality, daylight, comfort, and mindfulness.
- *Social equity* – Equality, gender equality, participation of people and countries, and equity in allocating resources, inclusiveness of people and generations, supporting vulnerable people such as empowerment of women, older and young people, giving disadvantaged groups a voice, no threat associated with food production, and globally responsible action in dealing with resources.
- *Economy* – Regenerative, circular, crowdfunding, and sharing economy; sustainable production and consumption; collaborative business approach; place-based economies at multiple scales, such as restorative enterprises; redesign business models with focus on selling products that create waste to provide services in closed-loop models; and energy unions.
- *Culture and community* – Address the social aspects of health, foster social cohesion and community identity, promote accessibility and integration such as re-integrated and enjoyable heritage, buildings, and local heritage as visual, social, cultural, and economic catalysts for the community, accessibility of cultural and historical places, and inclusiveness of rural communities.
- *Education and inspiration* – Enable and encourage stakeholder participation, bottom-up cultures and initiatives, whole-life education, increase awareness, such as encouraging pioneer movements such as permaculture, urban gardening, place making, and so on, giving a voice to different sectors and interests of society, continuous and intergenerational learning and feedback, cooperation, interaction, and interdisciplinary planning.
- *Environment and mobility* – Reduce carbon emissions caused by travel and transport, encourage walking and cycling, walkable and cyclable cities, rural–urban

balance such as reducing transport volumes at the building site, encouraging and enabling car sharing, e-car charging stations, bicycle parking spaces and cycle paths, and footpaths to schools and shops.

Further examination of available tools and certification schemes that support regenerative design approaches of a construction project can be found in Chapter 7.

6.2 The construction process

The construction process begins once the client initialises its procurement with specified documents outlining how the work will be performed, depending on the format of the contractual agreement and when the contractor (or contractors) acts as tenderer. The procurement process ends when the client has chosen and finalised an agreement with the contractor(s). The contractor then fulfils this contractual agreement during the construction phase when all the practical work is completed on a finished building on the site. The contractual work ends when the finished building is handed over to the client and is eventually followed with some complimentary commissioned work.

Procurement

Public and private clients of construction projects can procure material, contractors, and services with regards to sustainability to avoid the use of unsustainable products and services. This effort can include bulk procurement or minimum performance specifications for procurement rules. The goal is to increase the level of 'green' procurement, that is, procurement based on minimum specifications due to the standard ISO 20400, sustainable procurement (see below), or other requirements stated by the client, such as minimum energy efficiency, minimum embodied carbon, and other environmental standards, or due to client specified environmental building certifications. The specifications and documents required depend on the type of construction contracts stipulated by the client for the actual construction work.

Sustainable procurement ISO 20400

Sustainable procurement by ISO 20400: Sustainable procurement – Guidance, is an international standard that provides a process for making purchasing decisions for goods and services that is beneficial to both an organisation's needs and to society as a whole, while minimising the impact on the environment. Sustainable procurement is achieved by ensuring decent working conditions for the suppliers' employees that the actual products or services are, if possible, sustainable, and that socio-economic issues, such as inequality and poverty, are addressed.

The standard ISO 20400 provides guidelines to integrate sustainability into processes of procurement for an organisation, defining principles of sustainable procurement, including accountability, transparency, respect for human rights, and ethical behaviour. It highlights considerations such as risk management and priorities in establishing and implementing the stages of the procurement process, outlining requirements for integrating social responsibility. Sustainable procurement is a key aspect of social responsibility ISO 26000 (mentioned in Chapter 3, in Section 3.2) and is the basis of ISO 20400, containing the same principles and core subjects of human rights, labour practices, and fair business practices. Within the purchasing process of a construction project, the principles of social responsibility as described in ISO 26000 are integrated by following the guidance of ISO 20400.

ISO 20400 is aimed at organisations of all sizes in both the public and private sector and applies to purchasing decisions regarding building material and services, among many others. The standard provides guidelines (not requirements) and is not intended for certification purposes.

Construction contracts

There are several types of agreements, i.e. construction contracts, between the client and those who perform the actual construction work. Contractors may comprise a few or one main contractor with several subcontractors who agree with the client to perform the work based on the client's design and requirements and, of course, prevailing building codes and regulations. In a later stage of the procurement process, a few standard procedures of a contract are often modified to suit the client's needs and expectations and the specific needs of the actual construction project.

A letter of intent – an agreement of restricted scope that can serve as a precursor of the more formal standard options of contracting – can be used by the client to initiate an early start of the construction work by the chosen contractor. The scope of performing the work is limited, e.g. it can stipulate the amount of work, a time frame within which the work must be performed and completed, or limitations in the cost of the work to be conducted. This precursor – the letter of intent – is often followed by a more specialised formal contract for construction.

Most contemporary, formal construction contracts largely adhere to one of the following standard formats:

- *Stipulated or lump-sum contract*
  This standard format of a construction contract implies a complete set of the construction project documents prepared, reviewed, and quality checked by the design team before being issued for tendering. A critical component of the lump-sum contract approach is that if there are missing details, inconsistencies, or a lack of clarity in the documents, the total cost of the contractual work would most likely increase by additional costs due to these errors. When using the lump-sum contract approach, it is crucial to ensure that the intended documents are as complete and error-free as possible. The client must allow the design team enough time to ensure the accuracy of this quality check.
- *Value engineering*
  Value engineering is a process where the contractor can offer products, equipment, or construction techniques with some variance due to the contractor's interpretation of the project documents and specifications, but which is acceptable to the design team. Nevertheless, caution must be taken when a contractor submits a suggestion for value engineering. If the contractor suggests a specific measure concerning a component to save costs, it may result in increased cost for the client of another related component. The possibility of different outcomes should be explored and analysed throughout the intended performance in question. For example, a contractor may suggest that one large rooftop ventilation unit should

be replaced with two smaller units to reduce the cost and improve indoor climate control. However, this change could lead to a more expensive electrical circuit installation and extra costs when an additional roof opening and frame must be added. A reduction in one area can have a cascading effect and impact other areas. Suggested changes must be reviewed and analysed thoroughly to assess additional costs, including operating costs, durability over time, lifecycle cost, increased carbon emission, environmental impact, and increased amount of performance compared with the contractual time frame.

- *Cost of work plus fee contract*

  This type of contract stipulates a specific performance by the contractor specified by the client, with payment of the direct built-in costs plus a predetermined fee as reimbursement for indirect expenses. When the cost of the actual complete work is difficult to determine, no maximum limit in costs is provided for the contractual performance. This agreement is often used when the client must repair or prevent severe damage, often in an emergency situation, e.g. flooding, fire, or other weather-related situations when immediate actions of construction performance are needed. Another situation in which this form of contract may be used is when work must be done, but the detailed scope of the performance is unclear, such as deconstructing an older building with unclear structural and material content.

  Whether or not the client chooses a cost, plus a fee contract, largely depends on the quality, reputation, and integrity of the contractor. When the client is knowledgeable regarding a contractor, the performance and costs can be fairly accurate; however, conversely, an unfamiliar contractor could perform unnecessary or excessive billing for the work. The selection of a contractor is a key issue, as is defining the nature of the work without documents that specify the performance; usually, broadly descriptions are provided.

- *Cost of work plus fee with guaranteed maximum price contract*

  This is similar to above, however, the contractor's direct costs are built-in to the billing for their performance, plus a fee for reimbursing indirect expenses. The difference is a mutual agreement between the client and the contractor on a cap, or guarantee, on the maximum costs of the performance of the work. This could be used by a client who is eager to begin the construction work prior to the completion and review of the required documents and specifications, e.g. when the schedule is very tight to complete the final product, the building. The design team develops the documents and specifications to a point that a skilled and experienced contractor is able to estimate the efforts needed, both from the documents and the forthcoming performance required to accomplish the work. A contractor who is skilled and experienced in the particular nature of the actual construction project can reasonably estimate the total costs of the performance. Alternatively, a contractor not so well versed in the actual nature of the relevant construction performance could over-estimate the total costs, including costs for developing and completing the remainder of the necessary documents.

- *Construction management (CM) contract*

  A construction–management contract outlines an agreement where the construction manager acts as the client's agent, performing work on the client's behalf on construction-related matters. This differs from the previously described contracts, for which the relationship between the client and the contractor is opposite, where the contractor acts independently as an operator hired by a client to perform a construction project. The contractor negotiates and allocates specialised work to a subcontractor in the name of the contractor. Whereas, per a CM contract, after consultation with the client, the construction manager allocates the work to contractors in the name of the client.

The CM agreement essentially provides a professional service to accomplish a product – the construction project delivered by the contractor. Two basic types of construction manager exist: (1) a CM who provides all services included in the contract for a fee and receives payment based on a percentage of the final project costs and (2) a CM at risk, who provides all services included in the contract and receives payment based on a percentage of costs, with the final cost of the project guaranteed.

A construction manager contract provides plentiful services for the design, construction, and operational phases of a project. In the design phase, the CM and design team can monitor costs together as the design progresses, suggest information about available materials and costs locally, make use of the local experience with certain building systems and components, make use of specialists regarding sustainability and carbon emission matters, monitor and manage requirements concerning actual environmental building certificate systems, and develop progress schedules for both design documents and construction activities.

- *Design–build contract*

  The design–build contract integrates clients, contractors, architects, and engineers in a project delivery system with a more equitable distribution to all parties. Conversely, the lump-sum, competitively bid project carries risk when unqualified contractors are allowed to participate in the bidding process. Risk aversion within each party could easily result in claims, disputes, and litigation. The cost of work plus fee with a guaranteed maximum price contract approach provides a more equitable solution, based on actual documented costs, and caps the final cost of the project. The construction manager approach gains more favour in relation to its more client-oriented approach.

  The design–build contract allows the client, design team, and contractor to work in a team environment rather than as adversaries. This form of contract includes both the design and the construction components, and the client works with one entity, a design–build firm, instead of managing separate contracts with the design team and the builder for construction. The design–build team can be initiated in many ways:

  o A contractor with architect(s) on staff or, conversely, an architect with construction professionals on staff, forms the design–build team. In both cases, the project is performed within the organisation.

  o A contractor who has a relationship with a design firm teams up and offers services as a design–build team, generally as a joint venture or limited-liability corporation (LLC) with the design consultants as subcontractors.

  o An architect forms a joint venture or LLC and hires a contractor as the participant.

  o Partnering, a variant of LLC, enables the establishment of a cooperative approach with mutual objectives, making use of specific tools and techniques such as facilitated workshops, conflict resolution techniques, and continuous improvement techniques. This aims to build trust among the contracting parties, client, design team, and contractor. It can also enable stronger commitments to quality, sustainability, and safety performance, as well as to dispute resolutions, human resource management, innovation, and time and cost reductions. Partnering can be used as a variant during complex and customised projects with high uncertainty and long duration, coupled with severe time constraints.

Another standard usually attached to the chosen construction contract is known as 'General conditions to the contract for construction'. If available, depending on country and region, this document operates as the 'rules and regulations', statement. It defines the roles of the client, design team, and contractor through construction, listing the obligations and responsibilities of each, and elaborating on specific events that take place during the process. This document is often written through the cooperation of construction sector organisations.

Tendering process

Depending on the type of construction contract set by the client, different demands may be required of the contractor during the tendering process, i.e. basic requirements for the ability to tender and tendering requirements depending on specific client demands of the actual construction project. Demands for being a tenderer could be organisational, e.g. the requirements of quality and EMSs such as ISO 9001, ISO14001, or the European scheme EMAS, combined with ethical considerations such as ISO 26000 (Corporate Social Responsibility) or UN Global Impact (see also Chapter 3, in Section 3.2). Depending on the performance of the actual project, the demands of the tenderer could include certified knowledge or verified experience from the required certification systems (via the client).

Construction

Contractor organisation

As the requirements of the product – the building – increase in relation to sustainability and its forthcoming performance of sustainability, the requirements of the performer – the contractors – also increase regarding their work towards the final product. These increased requirements originate from new environment-adapted building codes as well as from stricter (public and private) client demands. Examples of contractor requirements could include the obligation to report the climate impact of the contractual work, with a focus on energy efficient working machines, the transport of materials, and material selection. Some projects entail making LCA estimations or calculations regarding carbon emissions throughout the building's life cycle. Some contractors using an EMS may have incorporated compliance obligations in its organisation, such as mandatory requirements of applicable laws and regulations together with the organisation's own voluntarily commitments, e.g. industrial or organisational standards, codes of practise, work environment commitments, agreements with community groups or NGO's, and so on. Codes of practise could contain routines to prevent bribes, fraud, inappropriate cooperation during procurement, carelessness or negligence in quality verification or work on the site, and so on.

Most of the contractor's organisational demands should be articulated in routines connected to the contractor's construction project plan for the specific construction work. These routines could contain specifications concerning:

- Onsite appointment of environmental responsibility
- Fulfilling sufficient education concerning environmental issues on site
- The establishment of an environmental programme for the project that includes how to manage the project concerning its environmental aspects
- The establishment of an environmental plan for the project, connected to the time schedule, that identifies various aspects and their occurrence, i.e. when according to the time schedule and where on the structure they occurs, and who is responsible for verifying their performance and how
- How to manage deviations from the project's environmental programme and plan, and how to communicate the deviations to the client
- How to manage unforeseen circumstances concerning environmental aspects during project performance
- Demands of lifecycle requirements concerning the supply chain of material selection and procurement of subcontractors
- Knowledge in applicable certification systems concerning sustainability requirements or regenerative outcomes (see more concerning certification systems in Chapter 7)

Site planning and construction performance

When beginning the contractual work, the contractor(s) should plan and optimise the layout of the provisional construction site based on necessity from an energy consumption and sustainability point of view. They must determine how to design workforce facilities and the site office based on energy efficiency, how to manage the delivery and storage of material, and how to plan the routes of material transportation and when deliveries occur. Methods of how to conduct the work must also be considered from a sustainability point of view, e.g. weather proofing, controlling dampness, minimising material consumption, sufficiently dehydrating moisture sensitive materials and processes (e.g. concrete casting), dehumidification processes for ensuring a healthy indoor environment, and so on. In addition, all package material from delivered building materials must be divided into different fractions for circularity matters (see more about circularity and the built environment in Chapter 2). For circularity reasons, it is also important to responsibly manage and sort all residuals from the contractor and subcontractor performances.

Supply-chain management

When transforming materials and labour into a complete project – a building – it is necessary to support the construction function by managing this process of transformation. The function of managing this supply chain is to integrate this transformation process entirely. Materials and components are combined with labour, subcontractors, information, technology, and capital during the process. The contractors' planning and scheduling system is essential when procuring materials and subcontractors effectively for the construction environment.

It is not only what the contractor is doing on the site that's important regarding sustainability. The contractor has to be aware of what happens during the entire supply chain process of selecting building materials and how procured subcontractors, and their subcontractors, function during their performance. The latter must be described in the tender documents for the subcontractors identified by the contractor.

Material selection

Depending on the contractual circumstances described above, the contractor selects and procures building materials mostly from the suppliers that the contractor uses on a regular basis. If the contractor uses EMS for its organisation, e.g. ISO 14001 or EMAS, they are required to consider how to procure material from a lifecycle perspective. Knowledge of the precautionary principle and substitution principle is a prerequisite.

According to the Rio Declaration (1992) principle 15, the precautionary principle is defined as follows:

> *In order to protect the environment, the precautionary approach shall be widely applied by States according to their capabilities. Where there are threats of serious or irreversible damage, lack of full scientific certainty shall not be used as a reason for postponing cost-effective measures to prevent environmental degradation.*

According to REACH (2022), the substitution principle is defined as a 'progressive substitution of the most dangerous chemicals … when suitable alternatives have been identified' and is used to stop dangerous or hazardous chemicals from building materials and components from affecting humans and the environment.

Several material selection tools can be used to identify chemical content that has been assessed as dangerous, both at the producer level and the building component level. Third-party environmental labelling systems can also be used for certain materials internationally and regionally. These labels approve the specific maximum content of chemicals that affect humans or the environment, including dangerous or hazardous emissions and using toxic constituents during production of the material. One important detail when selecting building material is having knowledge of chemical reactions between two or more seemingly harmless different materials when they are incorporated into one functional entity.

Various contractors have made a list of concerning materials and components to avoid because they contain specified dangerous chemicals. When uncertainty prevails, the contractor should use, as a guideline, the precautionary principle and the substitution principle. If verification of material or other criteria from building environmental certifications is required during the contractual work, knowledge of how to execute and verify these certification criteria is of great importance.

Subcontractors

Construction is project-driven and requires many skilled specialties; thus, a considerable amount of work is performed by specialist subcontractors. Electrical, ventilation, plumbing, mechanical, trucking, structural steel, painting, and roofing services are just some examples of work that is routinely subcontracted. Because the capabilities of subcontractors are very specialised, and due to the nature of their work and their role in the work environment, they are not as time sensitive to the design–built processes as the contractor. It is more a quantitative matter of cost to manage the selection of the most adaptable subcontractors into the project. In fact, subcontractors hold no contractual agreements directly with the client, i.e. no formal relationships exist with the client. The contractor is entirely responsible for the relationship with the subcontractor and for the subcontractor's performance regarding the client (see Figure 6.6).

As described previously, similar requirements apply to the agreement between contractor and subcontractor. This agreement should clearly state that those requirements are also valid for the subcontractor's subcontractor. Requirements similar to those between the client and the contractor regarding sufficient education in occurring environmental issues also apply to the subcontractor's performance. In their agreement with subcontractors, the contractor should demand an EMS that is similar to that used by their own organisation, i.e. requirements concerning material selection, auxiliary material, and working machines similar to those between the client and the contractor.

Handover and commissioning

The contractual agreement between the client and the contractor ends with the handover phase. By the handover of the finished construction project to the client, sufficient documentation of the performance conducted should be delivered to the client. Among other required documents, some relate to sustainability as follows:

- Sufficient declaration of built material and component content, e.g. in terms of construction product declarations, EPDs (Environmental Product Declaration), material passports, and so on.
- Documentation of built third-party environmental labelled product
- Documentation of built materials or products selected per the substitution principle and approved within the contractual work
- Documentation of built materials or products considered environmentally hazardous that were not substituted
- Verified environmental programme and plan, including corresponding verifications of performance
- Approved deviations, within the contractual work, from the environmental programme and plan, including corresponding verifications of performance

If agreements have been made or requirements have been stated in the contractual documents, or if it is an obligation within the actual building environmental certification, sufficient time and recourses for commissioning must be ensured by the contractor.

Building commissioning can be defined as testing and verifying equipment, e.g. HVAC- and other building systems fulfil the demands of the client and the codes and regulations regarding its intended performance. Commissioning is beneficial when reducing future maintenance costs, lowering energy consumption and carbon emission, and improving indoor air quality. But, significant energy and carbon savings can only be achieved during the design phase of the construction project, during which commissioning is a process and the commissioning agent collaborates with a team that includes the project's designers and mechanical engineers to enhance performance from a long-term perspective. It is important not only in commissioning the

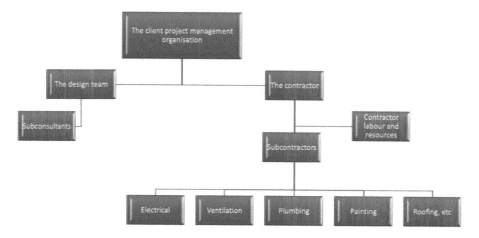

Figure 6.6 Relationship subcontractor – contractor - the client (Author original).

HVAC systems but should also especially take into consideration systems with high energy consumption, high carbon emissions, energy efficient systems, more complex systems, and systems affecting the indoor environmental quality.

Regenerative aspects of the construction phase

Procurement and contracting

The procurement phase of a regenerative construction project must correspond with all sustainability criteria to be fulfilled in conjunction with its regenerative scope. The contract should enforce that the contractor and subcontractors in all levels perform as stipulated by the client and that the activities they perform onsite reflect the level of requirements of the contractual work. Furthermore, this phase should also cover risk, responsibility, and rights allocation among the contracting parties in relation to the construction manager of the regenerative construction project. The risks should be allocated to the party who can best control them. All rights and obligations of contracting parties should be covered and should not overlap. All statements in the contracts should be clear to avoid misunderstandings.

LLC or partnering, mentioned earlier in this section, can support regenerative construction projects as it supports sustainability in construction projects. Partnering enables a cooperative approach among the contracting parties and facilitates a commitment to invest in green initiatives and reduce the footprint of the supply chain. As partnering offers a solution to improve the central control of managing construction that is inherently highly uncertain and complex, it could also support the dynamic, innovative, and iterative nature of regenerative construction. Partnering can serve as a tool for contractors to establish a supply chain with reliable subcontractors and suppliers with a similar working culture that complies with the regenerative requirements. Furthermore, a subcontractor and supplier team within the partnering cooperative can support the learning process and knowledge sharing required for the performance of regenerative construction. The suppliers and subcontractors in regenerative supply-chain management must maintain knowledge regarding agile construction project management to support the iterative nature of the design process. Such iterations must be stopped at a certain stage in the design process. If they occur during the construction phase, variations can result in increasing the environmental footprint or other cascading impacts. Further, it can lead to an increase the final cost (e.g. causing rework of completed performance).

Furthermore, regenerative construction contracts can encompass all sustainability criteria, i.e. environmental as well as social aspects of construction performance (e.g. ensuring ethical sourcing, prohibiting unethical sourcing activities, discrimination, and adhering to CSR or UN Global Impact commitments).

Standard contracts used, described above as 'General conditions to the contract for construction', as well as company-specific contracts, must be updated. The contract documents should include in-house guidelines, EMSs such as ISO 14001, social standards such as CSR or UN Global Impact, and the company's compliance obligations (see ISO 14001) or own code of conduct.

As contracts in all levels must encompass all sustainability requirements, the contract documents should include the site waste or recycling management plan, and the suppliers and subcontractors' requirements for providing their staff with sufficient adapted education on sustainability and regeneration principles to improve their performance. The contracts should enable effective control mechanisms (e.g. various inspections of the production sites of suppliers and of the sites for subcontractors' performed works, if not of the actual construction site) to ensure that the required regeneration performance is fulfilled. The contract should also require suppliers and subcontractors to fulfil stakeholder inquiries, external and internal, to detect and obtain feedback on their weak points regarding the environmental and social risks; furthermore, it should initialise a requirement for continual improvement of their part in the regenerative construction process. Effective contract administration throughout the construction management process of a regenerative project is essential.

Construction phase

In a traditional construction project, the client would appoint architects and consultants to design and project managers to select the contractor and, depending on the contractual form, the contractor would procure subcontractors and material suppliers and other services needed on site. However, this approach is fragmented and adversarial and does not enable regenerative construction. Early contracting procedures must be established in the project to facilitate an approach that enables regenerative construction performance. Contractor skills must be introduced in the early design process to bring regenerative sustainability and design in a performable construction project and to introduce cost efficiencies to the preconstruction phase. The earlier the contractor is appointed in the design process, the greater the benefits they can bring to the project. A great deal of project and sustainability value created in the design process of the project could vanish throughout the construction phase without close and focused project management. Accordingly, effective contract administration must be conducted throughout the regenerative construction project by project management. Documentation (e.g. progress reports, photos of progress of site works) must be carefully maintained. Effective and transparent communication should be maintained throughout the project management process so that miscommunication and claims can be resolved without turning into disputes. In the initial part of the construction process, education (such as health and safety training) should be provided to the subcontractors and suppliers regarding sufficient sustainable and regenerative construction processes and requirements as well as onsite circularity management. Supplier risks, environmental and social risks, and supplier improvement potential should be detected, and necessary precautions should be taken to ensure the smooth flow of the work.

The construction phase should begin with effective resources and site planning to support a smooth flow of material, information, equipment, and labour, and to reduce counterproductive activities such as waiting and transportation. Resources that are shared must be coordinated. Effective site planning for regenerative construction projects affects the sustainability performance of the construction work as well as health and safety matters, e.g. failing to weather-proof material could cause undesirable waste. Effective site waste management and planning requires a well-structured management plan that outlines how to reuse or recirculate, agreements between contractors and subcontractors to determine who is responsible for managing the waste on site, procuring waste companies that offer beneficial waste segregation systems for recycling and disposal, and the appointment of a designated waste and recycling manager facilitate the delivery and storage of materials.

To establish sufficient and necessary precautionary measures, control activities should be conducted throughout the construction phase to identify and act on any deviances and to take necessary precautions in relation to time. Performed work needs to be controlled and comply with the contractual design documents and the specified control mechanism (e.g. various inspections, quality verifications, and visits to the site). Control activities of precautionary measures concerning embodied carbon and carbon emission should be conducted on the building product and supply-chain levels, complying with contractual requirements and the contractor's own compliance obligations. The suppliers and subcontractors should be educated on how to reduce their carbon impact and other potential sustainability deviances of their performance.

Handover and commissioning

The handover documents of the regenerative construction project should be provided to the client by the main contractor when the contractual stipulated performance is complete. These documents include operator manuals, maintenance manuals, spare parts catalogues, as-built drawings, and operational staff educational documents.

The operation manuals should include information on the supposed time between failures and the supposed time of repair. To understand supposed time of failure of construction systems, each system must be analysed to determine its failure and repair frequencies, i.e. its maintenance frequency. Supposed time to repair is the duration required for the repair in case of a failure, i.e. the necessary maintenance duration. These estimations of durations must be considered for the entire operation life of the regenerative construction project and its systems. These maintenance durations will affect the operational cost and revenue during the building's life cycle; an effective planned maintenance system is crucial to minimise unexpected repair measures due to unexpected failure. Planned maintenance system documents should be submitted to the client by the main contractor. Preparation of the planned maintenance system should include:

- Each specific piece of equipment and its expected durability
- System component and maintenance includes
  - o the time schedule when each specific maintenance should take place
  - o information on capacity
  - o required spare parts
  - o required duration of work hours
  - o qualification of maintenance personnel
- Documentation of vital equipment and systems considering operational requirements (especially for certain building types, e.g. hospitals, etc.)
- Equipment-specific standards
- Preparation of as-built or record drawings documents, including operator manuals, maintenance manuals, spare parts catalogues, and training documents for the regenerative construction project
- Education and training for operators, maintenance staff, and instructor documentation, and training warranty from the contractor. Sufficient education should be provided specifically to operators, maintenance staff, and instructors before the commissioning phase, at the end of the construction process. Sufficient verifications of the education requirements should be included in the handover.

6.3 Operation and maintenance process

Most of the issues concerning operation and maintenance (O&M) of the existing buildings can be found in Chapter 1 in Section 1.6, where issues can be found, such as building renovation passports, minimum energy performance standards, and key actions to achieve sustainability in the building stock. This section focuses on the operation and maintenance stage of a building's life cycle. It begins by describing key actions concerning O&M and follows with a section on a concept called 'Soft Landings', to facilitate the handover of a construction project to the client's operation management. This section also includes touches upon how to determine if everything works as intended by doing a post occupancy evaluation. The next section addresses how the sustainability relation between the client and the buildings end user, the occupant, can be solved through 'Green Leasing'. The subsequent section defines operation management or facility management with a sustainability concept, sustainable facility management, SFM, and provides definitions and descriptions of green maintenance and green maintainability. Lastly, various aspects of regenerative O&M are discussed.

Key actions of the operation and maintenance of a building

Although the construction project delivery of zero carbon emission, efficient, and resilient new or renovated buildings is essential, it is equally important to ensure that buildings are operated and maintained efficiently and as intended. Behavioural and operational management impacts the carbon emissions and energy performance of a building. Globally, key stakeholders to set this direction are those who can influence or demand (e.g. the end users) how a building performs or how it should perform, and those who deliver the supposed performance through adequate energy efficient, and zero carbon emission, operation. In addition, supporting functions exist through research, funding, educating, and enablers of technology availability. Various key factors of how to reach such operation and maintenance are listed below:

- *Benchmarking* – By tracking own, evaluating, and comparing other buildings' performance, operation management can make sustainability-verified investment decisions when improving the overall performance of the building operations. Through continuous improvements, this evaluation process could also support energy performance, certification, and building passport targets.
- *Certification for operational performance* – Building sustainability certification and energy labelling can be used to verify the performance and confirm or refine the performance requirements. The certification may be linked to the benchmarking above. The sustainability certificate or energy label enables the verification of documentation and adequate information sharing for occupants and relevant financial decisions. The certification of operational performance could also allow adequate formulation concerning green lease agreements, an agreement between a landlord and a tenant to enable the tenant to use and operate the building in a sustainable direction (see the section on green leasing below).
- *Building passports* – Building passports can be used to track information about the building, such as materials, systems, energy use, maintenance measures, repair, renovations, and other relevant information for improving the data for decision-making processes. For more about building passports, see the section 'The big challenge of existing buildings' in Chapter 1.
- *Energy audits* – Regularity of energy audits is important to assess the opportunities of measures to save energy and reduce carbon emissions, which is especially important in buildings with high energy consumption. An energy management system monitors the energy consumption of systems, components, and the building as a whole with the aim to identify anomalies and understand energy consumption trends. An energy management system can consist of a network of digital energy meters or sensors, or a simple smart meter.
- *Incentives* – Non-financial incentives, such as expedited permits or the allowance of increased floor area, could promote sustainability in building operation. Financial incentives can be used to support the very best of performed buildings (see more below). Both should be linked to building sustainability certifications.
- *Maintenance tools* – Operation and maintenance (O&M) manuals can temporally and actively support the maintenance of buildings with schedules for specific periodic maintenance measures (e.g. replacement of air intake filters). O&M manuals of a system should be provided at the construction project handover or new installation handover and be actively used by the building operator management in its digital building management system.
- A building management system can be a full-scale building tool digitally connected with multiple systems detecting and learning to improve the operation and management of a building. It can also entail simple controlling devices that manage independent technology within a building.
- *Audit tools* – Building sustainability or energy labelling audits energy and/or sustainability audits provide an opportunity for optimisation and systematically

check systems to identify priority maintenance and retrofit measures. Tools for auditing (e.g. software, sensors, and thermal cameras) can reduce the cost of a single audit and, accordingly, an opportunity to improve the rate of annual audits.

- *Sensors and controls* – Sensors and controls are fundamental to a smart O&M audit, and to operational energy- and building management. Control systems can range from fully digital centralised systems to simpler programmable devices, such as thermostats. Sensors and controls incorporate machine learning to understand occupant preferences and to optimise system settings based on internal and external conditions.

Different finance solutions can enable increased action towards zero carbon emissions, and efficient and resilient building operations. Examples of possible solutions for financing and promoting building operations sustainability outcomes can include credit lines of sustainability funding delivered through banks for a specific purpose, such as sustainability certification or development projects. It can also be through green bonds, i.e. bonds used to bundle funding associated with sustainability and regenerative development, or through direct governmental funding by tax-reductions concerning sustainable products and services. Other options include community financing or crowdfunding, collective funding from many people connected locally or through a call for funding, or, alternatively, leasing solutions, i.e. the ability to use sustainability-based products on a rental basis to reduce capital expenditure.

Information combined with education or capacity building activities can increase overall awareness, improve the decision-making process, and encourage sustainability choices. Educating or training professionals who work directly with operation management can enable increased ability, resources, and capacity to deliver sustainability in building operations, e.g. providing training programmes for service and product providers of buildings operations regarding how to implement cost-effective sustainability operational measures. Later, professionals can be educated about how to comply with sustainability policies for new or existing buildings, enhancement programmes for efficient operation and maintenance, or retrofitting and enabling certification or accreditation for operation professionals. A general commitment can also be made to promote campaigns or information concerning operation measures of buildings that owners or occupants can implement, and how to generally access appropriate funding for sustainability measures.

When operating buildings with regards to sustainability, multiple holistic benefits should be achieved, many of which are also linked to the UN Sustainable Development Goals, especially with SDG 7 (affordable and clean energy) and SDG 13 (climate action). Environmental benefits include the reduction of carbon emissions through lowered energy consumption, saving energy through efficient use of building systems, less strain on the public grid due to reduced energy use, reduction of high energy peak loads due to optimised operational energy use, and enhanced air quality due to the reduction of air pollution from sustainability operation. Economic benefits include increased resources for alternative investment options due to lowered operational costs; improved occupant productivity due to increased thermal, light, and acoustic comfort; greater demand for skilled operation service staff due to increasing efforts of sustainability in building (and, thus, an increased employment rate). From a social point of view, the sustainable operation of a building can increase health and well-being by improving physical and mental health through continuous commissioning with increased thermal, light, and acoustic comfort. In addition, safety can be improved through the prevention of system failure and increased and optimised sustainability operation and maintenance can lower the total operational cost and promote poverty alleviation.

Soft landings and post-occupancy evaluation

It is generally understood that matters of sustainability, energy efficiency, and the overall performance of new and existing buildings needs to improve radically in the coming few decades. Thus, the construction industry and its clients do not typically use existing structures to reliably make this improvement. Findings reveal vast gaps between client and designer expectations and actual delivered performance. There are many reasons for the lack of required and expected performance results, such as a lack of knowledge regarding occupant use and management of buildings and equipment, the need for careful attention and continual commissioning for complex and technology-intense operation environments, and systems that are too complicated to manage and operate efficiently.

To reduce or close these gaps between performance expectations and achieved level of performance, it is important to focus on technical and quantitative input, but also on in-use performance through considerations made during the very early design strategies. One method of addressing this gap is by applying the concept of Soft Landings, which connects the client's key requirement of performance to the final product – the building – and its long-term operational performance. Throughout the entire construction project process – from the early inception phase to final delivery and beyond delivery of the project – the concept ensures that the focus remains on these key requirements. The client objective could involve quantitative matters, such as maximum energy performance, or qualitative matters, such as user satisfaction (see below regarding post-occupancy evaluation, POE). This requires all stakeholders involved in the development of the construction project to share risk and responsibility, even after project completion. The Soft Landings concept comprises six main phases in the construction process:

- *Inception and briefing* – Establishing in-use performance criteria and performance outcomes and assigning responsibilities of fulfilment to the different criteria
  - o Reviewing project team members' experience from prior similar projects to facilitate better and more beneficial levels of target requirements. It is important to include the client's operation management staff in this review
  - o Defining how to measure and verify the target requirements
  - o Planning for intermediate and continual evaluation and realty checks, e.g. through workshops, to maintain team member engagement and target development throughout the process
  - o Establishing gateways for fixed decisions and follow-ups for target requirements
  - o Defining tender documents with appropriate commitments for all different and involved parties
- *Design* – Controlling the target requirements during the design phase through project management
  - o Procuring consultants and contractors with commitments throughout the entire construction process and beyond project delivery, i.e. design, construction,

handover, and commissioning, into the initial phase of the operation of the building

- o Reviewing experience from prior similar projects to facilitate the buildability, manageability, and usability of delivered buildings
- o Making advanced simulations of design proposals to verify possible outcomes during operation
- o Organising design reviews, preferably with third party reviewers, e.g. an impartial and independent expert team, to facilitate and highlight favourable design ideas and identify problems or possible errors in the design work. It is especially important to involve people with experience in building usage and operation
- *Construction* – Controlling the target requirements through the construction phase
  - o Continuing with advanced simulations as follow-ups to ensure that deviations during construction are meeting the target requirements
  - o Complementing in-use documents when similar solutions and equipment are chosen, other than those that were prescribed, e.g. when using the substitution principle
- *Pre-handover* – Ensuring that the handover and commissioning process are operable
  - o Organising routines of energy and environmental logging and verifications
  - o Organising readiness for building handover to the client and its operation management (as well as to the occupants)
  - o Commissioning document accuracy and programming for post-commissioning and fine-tuning
  - o Ensuring appropriate maintenance agreements and contracting
  - o Ensuring that operation and maintenance staff receive appropriate education regarding in-built systems and ensuring that the education is undertaken well before handover
  - o Coordinated migration planning of the occupants with project team members
  - o Planning and assuring in-site facilities for the aftercare team
  - o Compiling a guide for occupants and a technical guide for operators to ensure the smooth transition of intention and function of building facilities
  - o Checking, reviewing, and signing operation and maintenance documentation with the facility management
- *Initial aftercare* – Intention to help the occupants to understand the building and the operators to understand the systems (lasts about four to six weeks after handover)
  - o Confirming who (from the project team) will attend as members of the aftercare team onsite
  - o Confirming with the client or the occupant which in-site facilities are appropriate for the project team's aftercare team
  - o Organising the introduction of guidance regarding the building and its function to the occupants by the aftercare team
  - o Organising the introduction of technical guidance regarding the building and its systems to the operators by the aftercare team
  - o Initialising regular communication via the aftercare team, e.g. newsletters regarding operational progress, and informal walkthroughs to investigate occupant usage and prevent initial emergencies through observation and dialogue with the occupants

- *Extended aftercare and POE –* Regular visits by the aftercare team to monitor progress and feedback from occupants and operators for three years after handover (more regular visits during the first year and fewer regular visits after year one).
  - o Organising regular aftercare meetings with the aftercare team to review progress and feedback from occupants and operators – every three or four months during the first year and approximately one every six months after one year
  - o Logging environmental and energy performance via the operators to verify the true performance
  - o Fine tuning the systems in relation to seasonal variations via the operators and recording these fine tunings and usage changes into building logbooks and, if necessary, into the operation and maintenance manuals
  - o Initiating and organising POEs (see below), one after a full season variation (about 12 months) and one during year three
  - o Gradually decreasing the frequency of the aftercare team's regular communication and walkthroughs, which cease before the end of year three
  - o Organising and implementing a systems review (about once every six months) and an end-of-year review to verify and compare real performance with intended performance and adjusting to reach or achieve just below the intended performance. The last review in year three should be conclusive and include lessons learned regarding occupant, operator, and aftercare team experiences.
  - o Soft Landings is not an alternative to and does not compete with sustainability methods and tools, such as building certification schemes. Instead, it is a concept meant to compliment the other tools and methods because of its commitment to performance evaluation, graduated handover, and post-delivery follow-ups (including POE).

Post-Occupancy Evaluation

POE can be undertaken when the building is occupied to determine whether the construction project objectives and the sustainability outcomes stated in the objectives or in the sustainable building certification scheme, or later design targets for the building systems, have been achieved. The three progressive levels of POEs include (1) Light touch POE – a simple but meaningful rapid evaluation undertaken post occupancy, before all the contractual works, i.e. commissioning and extended handover, is complete. The information from this POE may not reflect the final building performance due to all seasonal commissioning being incomplete or other building systems not being fully in-bedded. However, it can provide useful insights for the client and serve as feedback for other projects. (2) Diagnostic POE – includes feedback from the Light touch POE and can identify a need for a more detailed evaluation. This could be undertaken by independent evaluators during the second year of occupation, verifying performance and reviewing any issues discovered, including those identified in the Light touch POE. (3) Detailed (forensic) POE – includes survey investigations, if necessary, by independent evaluators, to identify and, where possible, resolve any significant and persistent performance matters. The POE can start at any time but should ideally be completed by the end of the third year of occupation (see also the scheme 'Soft Landings').

There are numerous methods to accomplish a POE on the market, and the standard procedure differs depending on where the building is situated. A POE can be assessed quantitatively by technical schemes or with more or less qualitative results using survey or interview methods. A mixture of methods is often used to accomplish a POE, e.g. walk-through surveys, spot checks using small mobile instruments, and informal discussions with occupants and management.

Green leasing

The previous sections derive from client and operation management points of view and the commitment of sustainability in operation and maintenance, either by the client or the operation management. Regarding the end users – the occupants – and their commitment to 'user' sustainability of the facility they use, one solution could be an agreement, called a Green lease, that entails the requirements of such commitment. A Green lease is a physical agreement in print that is a legal document between the contractual parties – a landlord (here as client) and a tenant (here as occupant). Green leasing, however, describes a process to create and form an agreement with a content of sustainability commitment beneficial for both parties, but this process will not necessarily end up with a Green lease. The Green leasing process includes how to let out a building with sustainability performance in mind, with or without a Green lease, and non-sustainable practices of operation that are not compatible with the concept. Compared to standard leasing, Green leasing stipulates that the occupant(s) lease the building with considerations of the environmental impact, such as recycling, green cleaning practises, and occupant behaviour or involvement in operational practice. The development and increased use of building sustainability certifications by clients, stricter regulations, and changing commercial practise entails end users to increase their focus on their practice of using a building, whether it is attached with a Green lease or not. This increased focus implies a change in the approach to sustainability in commercial and non-commercial buildings and who and how stakeholders can be involved.

The client is the primal stakeholder responsible for building operation and its costs, even if operation management service enterprises are contracted to operate and maintain it, this is the base of a gross lease. In some cases, the occupant has taken over responsibility for operation, repair, and costs due to a significantly decreased rental cost; this is called a net lease. Hence, the net lease is considered to be occupant-incentivised, e.g. by occupant action to decreased energy usage. Another variant of basic lease agreements is when the occupant is responsible for all the costs related to the building they occupy and the net rental costs. This variant is often divided into three parts: real estate taxes, net building insurance, and net common area operation and maintenance. A Green lease often includes the occupant's portion of base rent obligations as well as the sustainability clauses and utility responsibility, aiming for a positive environmental impact of the building. Such a lease agreement could mandate the occupants in its clauses from a sustainability point of view, e.g. to reduce energy, co-invest with the client in the improvement of the building's infrastructure, or procure sufficient operation services.

One attempt to establish clause types and qualities in order to turn a standard lease into a Green Lease has been undertaken in Australia by the Better Building Partnership in the 'Leasing Standard Template Clauses' in 2016. The categories of clauses are divided into four sections: (1) Cooperation and Works includes environmental initiatives, sustainable upgrade works, sustainable management collaborations, premise designs for sustainability, and waste management from works and social initiatives. (2) Management and Consumption includes management areas such as energy, water, waste, and indoor environments, and follows with sustainable utilities, including cleaning, transport, and sustainable procurement. This section is more relevant to operation management. (3) Reporting and Standards includes information sharing, performance ratings, design/development ratings, performance standards, comfort, and metering and monitoring. This section of clauses is relevant to both occupants and operation management. (4) Compliance and Costs includes dispute resolution, assignment, and rent review, and concerns the agreement itself. These four sections of clauses do not comprise a comprehensive list of requirements but offer a base from which to improve, update, or develop new sets of clauses, depending on client objectives, occupant demands, or regulatory causes.

However, the relationship between the client and the occupant is not the only factor to consider regarding a Green lease agreement. The operation management enterprises, or facility managers, are increasing their efforts and highlighting the importance of managing towards more sustainability outcomes due to higher demands from the clients, society, and the facility management market.

Sustainable facility management and green maintainability

The operation of buildings is often the responsibility of the Facility Management (FM) profession or enterprise, either as in-house or as a contractual service to the client. Facility Management, in particular, is able to contribute to a sustainable interdependency between the built environment, the natural environment, and the client organisation business environment. The FM is defined by the international standard ISO 41011:2107 as an 'organisational function which integrates people, place and process within the built environment with the purpose of improving the quality of life of people and the productivity of the core business'. This standard also establishes the core elements of FM practices, such as to:

- Improve quality, productivity, and financial performance
- Enhance sustainability and reduce negative environmental impact
- Develop functional and motivating work environments
- Maintain regulatory compliance and provide safe workplaces
- Optimise lifecycle performance and costs
- Improve resilience and relevance
- Project an organisation's identity and image more successfully

However, with a greater focus on sustainability matters in the built environment, as well as from the client and occupant point of view, the intention and demands of FM sustainability performance have also transitioned to a sustainable facility management (SFM) enterprise. Also, depending on the developing context of green leasing and sustainable building certifications, FM performance is heading towards adopting SFM performance. SFM has a lot of similarities due to the standard definitions of FM; however, environmental impact is kept in mind when decisions concerning procurement, implementation, and strategy are made. SFM differs from FM through the consideration of not only core business and support functions but also in relation to local and global society as well as to the climate and the ecosystem and its ecosystem services. SFM extends the scope from a single building to the building environment of the site environment and, consequently, the entire built environment, as competencies and knowledge lead to the implementation of new and sustainable technologies and practices in the built environment.

Green maintainability

Maintenance during a building's life cycle can be defined as activities carried out to sustain the performance, usage, and value of the building by maintaining, repairing, retrofitting, or upgrading the components, devices, services, and grounds, to an up-to-date standard. When including sustainability matters in the concept, green maintenance can be defined as maintenance that adopts a sustainable management approach in maintenance processes, methods, and materials with minimised negative environmental impacts and improved health and safety of maintenance personnel.

Conversely, maintainability is defined as the attainment of optimum performance during the entire life cycle with lower lifecycle cost. This also involves the input of maintenance knowledge during the design process to ensure effective and efficient maintenance. When considering sustainability during the design process, green maintainability requires maximising resource and energy efficiency performances and minimising lifecycle costs, embodied carbon, and energy and material consumption. This requires a base of SFM input, mentioned above. Furthermore, the mentioned lifecycle costs include the maintainability costs, i.e. direct maintenance costs and the costs from environmental impact, waste, and emissions generated during maintenance activities.

When implementing green maintenance, a lack of green maintainability considerations during the design stage is the biggest impediment. As a result, sustainability considerations taken during the design process do not necessarily lead to a sustainability constructed building due to a lack of sustainability practices during the operation and maintenance stages. Sustainability and green maintainability considerations should be applied throughout the entire life cycle of the building to bridge the knowledge gap between designers and the client's operation and maintenance management and to achieve resource and energy efficient and effective O&M. Thus, participatory design and user involvement are essential to the sustainability and green maintainability of the buildings. Unfortunately, no specific rating tool for green maintainability yet exists on the market; only a few guidelines are available.

Regenerative aspects of operation and maintenance

The process of a regenerative construction project must be considered as one that continually evolves over time. Regenerative development during the design process and the construction stage does not end with the delivery of the final drawings and approvals, nor with the construction and handover of a project. As these projects are like living organisms, their regeneration performance should be kept continuously up to date with contemporary technologies and, considering the dynamics of nature, comprehension of the state of the art of sustainability today will change even over the next decade. Regenerative development implies continually maintaining its adaptive capacity of unpredictably and capacity to renew its knowledge. Therefore, throughout the building's life cycle, contemporary innovations and new equipment and systems should be analysed and thoroughly examined and the impact on the ecosystem should be monitored through ecosystem analyses (see Chapter 4). If the existing equipment is outdated compared to the new, a schedule for exchanging or upgrading should be carried out. The SFM contractor should be responsible and defined in the contract between the client and the SFM contractor. Depending on the circumstances and contractual requirements, e.g. partnering or Soft Landings, and so on, the client or SFM contractor can transfer this responsibility to the construction contractor due to their knowledge of construction performance, installation skills, or familiarity with the equipment. If the installation of the equipment, for which a space and infrastructure has been constructed beforehand, can be carried out in the post-construction phase, the relevant supplier can complete the installation. Depending on the conditions of the contract, this can be coordinated by the client, the SFM contractor, or the original construction contractor. Both of these concepts can be supported by the design and construction allocated during the design process.

Maintenance and repair, based on the maintenance schedule defined in the contract between the client and the construction contractor, can be coordinated by the contractor with the SFM or a relevant supplier, as constituted in the contract conditions. Moreover, a framework is needed for integrating social and cultural aspects into SFM practice, especially small and medium FM enterprises, to perform the regenerative development of the O&M.

To turn SFM into regenerative facility management, various targets should be established for social, economic, and ecological regenerative development.

The latter ecological targets could include:

- Reduction of resources with a focus on circular economy
- Usage of recyclable building materials
- Consideration of disassembly and re-use of material
- Reduction of energy consumption, embodied carbon, and usage of renewable energy
- Reduction of space requirements
- Safeguarding the ability to maintain, reuse, and deconstruct buildings
- Substitution of dangerous and hazardous materials impacting people, or the environment, based on the precautionary principle

Economic targets for regenerative facility management could include:

- Building optimised space for more efficient usage, using BIM and virtual realities digital technology to monitor effectiveness
- Optimisation of lifecycle costs involving different stakeholders to design, construct, handover, use, maintain, and re-develop the buildings in the long-term, i.e. circular economy perspective
- Facilitating the most efficient operation and maintenance management methods enhanced by digitalisation, ecosystem practices, and responsible procurement
- Using green bonds and crowdfunding as the basis of financial management

Social targets to facilitate regenerative development could include:

- Supply of a balanced number of buildings for work and life, developing mixed-use and hybrid facilities in the context of urban regeneration

- Physical and psycho-social well-being, together with health, safety, and security requirements
- Identification of different social groups and social impacts – resilient buildings and resilient neighbourhoods integrate different social groups and provide synergy
- Communication of regenerative values for users – increases awareness of regenerative actions

It is important to identify the impact and opportunities of a regenerative approach to both construction and operations, in addition to a regenerative use of buildings. It is also essential to continually optimise the benefits for the environment and users and to ensure that the initial quality is maintained or enhanced. Per Peretti and Durhmann's (2019) conclusion, the steps towards a regenerative Facility Management could include transitioning:

- From scheduled maintenance towards on-demand maintenance
- From recycling only towards self-sufficient solutions
- From a linear economy towards a future circular economy
- From human intelligence towards artificial intelligence
- From the passive user towards the active user
- From monitoring single indicators towards monitoring integrated indicators of co-operation and connectedness
- From maintenance by a service provider towards maintenance by the user and prosumer

Bibliography

Asmone, A.S. and Chew, M.Y.L. (2016) *Sustainable facilities management and the requisite for green maintainability*, Paper presented at the Challenges & Opportunities for Facilities Management in AEC, Singapore, www.researchgate.net/profile/Ashan_Asmone/publication/308777626_Sustainable_facilities_management_and_the_requisite_for_green_maintainability/links/5811b61508ae009606be8b35.pdf, access 2022-02-016.

Asmone, A.S. and Chew, M.Y.L. (2018) *Green maintainability conceptual framework*, conference paper, 1st International Conference on Construction Futures, Wolverhampton, UK, 2018.

Attia, S. (2018) *Regenerative and Positive Impact Architecture – Learning from Case Studies*, SpringerBriefs in Energy, Springer Nature, Cham, Switzerland, 2018.

Benton, W.C. Jr. and McHenry, L.F. (2010) *Construction Purchasing & Supply Chain Management*, McGraw-Hill Companies, New York, NY, 2010.

Better Buildings Partnership, BBP (2016) BBP Standard Leasing Template, https://s3.ap-southeast-2.amazonaws.com/cdn.sydneybetterbuildings.com.au/assets/2016/09/BBP-Model-Lease-Clauses.pdf, access 2022-02-16.

Biomimicry Institute (2021) www.biomimicry.org, access 2021-09-16.

BSRIA (2014) *The Soft Landings Framework – For Better Briefing, Design, Handover and Building Performance in-Use, BSRIA and Authored by the Usable Building Trust*, BSRA, BG 54/2014, https://www.bsria.com/uk/product/XBYWwn/how_to_procure_soft_landings_2nd_edition_bg_452014_a15d25e1/, access 2022-08-28.

Chew, M.Y.L. et al. (2017) *Developing a Research Framework for the Green Maintainability of Buildings*, Facilities 2017, 35(1/2), 39–63.

Clark, G. and Tait, A. (eds.) (2019) *RIBA Sustainable Outcomes Guide*, Royal Institute of British Architects, RIBA, London, UK, 2019.

Cohen, R. and Ratcliffe, S. (2019) *Soft Landings and Design for Performance*, BSRIA, BG 76 /2019, Bracknell, UK, 2019.

Collins, D.A. (2019) *Green Leasing – A study of barriers and drivers for Green Leased office in Norway*, Doctoral Thesis at NTNU, 2019:150, Norwegian University of Science and Technology, Trondheim, Norway, 2019.

Day, C. and William, J.G. (2020) *Living Architecture, Living Cities: Soul-Nourishing Sustainability*, Routledge, New York, NY, 2020.

Elzarka, H.M. (2009) *Issues in Building Commissioning*, The Journal of the American Institute of Constructors 2009, 33(2), 18–29.

EN ISO 14001:2015 (2015) *Environmental Management Systems: Requirements with Guidance for Use*, CEN European Committee for Standardization, Brussels, Belgium, 2015.

Eriksson, P.-E. (2010) *Partnering: What Is It, When Should It Be Used, and How Should It Be Implemented?* Construction Management and Economics 2010, 28(9), 905–917.

European Commission (2022) REACH, https://ec.europa.eu/environment/chemicals/reach/ reach_en.htm, European Commission – Chemicals, access 2022-01-27.

GlobalABC/IEA/UNEP (Global Alliance for Buildings and Construction, International Energy Agency, and the United Nations Environment Programme) (2020) *GlobalABC Roadmap for Buildings and Construction: Towards a Zero-Emission, Efficient and Resilient Buildings and Construction Sector*, IEA, Paris, 2020.

Hansson, B. et al. (2017) *Byggledning – Produktion (Construction Management – Construction Phase)* (Swedish only), Studentlitteratur, Lund, Sweden, 2017.

Haselsteiner, E. et al. (2021) *Drivers and Barriers Leading to a Successful Paradigm Shift toward Regenerative Neighborhoods*, Sustainability 2021, 13, 5179.

ISO (2017) *ISO 20400 Sustainable Procurement – Guidance*, International Organization for Standardization, Geneva, Switzerland, 2017.

Jamei, E. and Vrcelj, Z. (2021) *Biomimicry and the Built Environment, Learning from Nature's Solutions*, Applied Science 2021, 11, 7514.

Levy, S.M. (2010) *Construction Process Planning and Management – An Owners Guide to Successful Projects*, Butterworth-Heinemann, Burlington, MA, 2010.

Loftness, V. (ed.) (2020) *Sustainable Built Environment, Encyclopedia of Sustainability Science and Technology Series*, 2nd edition, Springer Science+Business Media, New York, NY, 2020.

Nkandu, M.I. and Alibaba, H.Z. (2018) *Biomimicry as an Alternative Approach to Sustainability*, Architecture Research 2018, 8(1), 1–11.

Pawlyn, M. (2016) *Biomimicry in Architecture*, 2nd edition, RIBA Publishing, London, UK, 2016.

Pedersen Zari, M. (2015) *Ecosystem Services Analysis: Mimicking Ecosystem Services for Regenerative Urban Design*, International Journal of Sustainable Built Environment 2015, 4, 145–157.

Pedersen Zari, M. (2017) *Biomimetic Urban Design: Ecosystem Service Provision of Water and Energy*, Buildings 2017, 7, 21.

Pedersen Zari, M. (2018) *Regenerative Urban Design and Ecosystem Biomimicry, Routledge Research in Sustainable Urbanism, Earthscan from Routledge*, Routledge, Abingdon, UK, 2018.

PEEB (2020) *Better design for cool buildings – How improved building design can reduce the massive need for space cooling in hot climates*, PEEB working paper August 2020, Programme for energy efficiency in buildings, PEEB, Paris, France, 2020.

Peretti, G. and Druhmann, C.D. (eds.) (2019) *Regenerative Construction and Operation, REthinking Sustainability TOwards a Regenerative Economy*, RESTORE, Working group three publication, Eurac research, Bozen, Germany, 2019.

Phytofilter Technologies (2016) Phyto-purifying systems, Whole Building Design Guide (WDBG), National Institute of Building Sciences, Washington, https://www.wbdg.org/ resources/phyto-purification-systems, access 2022-01-09.

Sertyesilisik, B. (2017) *A Preliminary Study on the Regenerative Construction Project Management Concept for Enhancing Sustainability Performance of the Construction Industry*, International Journal of Construction Management 2017, 17(4), 293–309.

UNEP (2020) *2020 Global Status Report for Buildings and Construction: Towards a Zero-emission, Efficient and Resilient Buildings and Construction Sector*, United Nations Environment Programme Nairobi, Kenya, 2020.

UNEP (2021) *2021 Global Status Report for Buildings and Construction: Towards a Zero-emission, Efficient and Resilient Buildings and Construction Sector*, United Nations Environment Programme Nairobi, Kenya, 2021.

United Nations, General Assembly (1992) Rio Declaration on Environment and Development, A/Conf. 151/26 (Vol 1) Report of the United Nations Conference on Environment and Development, Annex 1.

WHO (2018) WHO Housing and health guidelines, https://www.who.int/publications/i/item/9789241550376, access 2022-01-04.

7 Verify – assessment, rating, and certification

7.1 Reducing carbon emissions

The scenarios established by the United Nations Intergovernmental Panel on Climate Change to reach the goal of a maximum of 1.5°of global warming by the end of the 21st century include reaching zero carbon emissions in the construction sector by 2050. This applies to both new and existing building stocks. Regarding cities, a considerable decrease in carbon emissions can be mitigated by reducing and changing energy and material consumption as well as by enhancing carbon uptake and storage in the urban environment. This can be achieved within and outside urban boundaries through supply chains that can also promote beneficial cascading effects. One mitigation strategy is to enhance carbon uptake and storage through bio-based construction materials, permeable surfaces, green roofs, tree planting, green spaces, and water surfaces such as rivers, ponds, and lakes. Thus, to enhance a decrease in carbon emission, nature-based solutions or ecosystem-based approaches in the built environment should be increased.

Carbon emissions in construction

Activities in the construction process generate a major flow of materials in every country and constitute approximately one-third of global material consumption and waste generation. Carbon emissions and energy consumption are linked to every phase of the life cycle of materials and products, from extraction or harvesting to manufacturing, transport, construction, usage, deconstruction, and disassembly. Materials such as steel, cement bricks, and non-certified wood (a deforestation issue) are major building material emitters of carbon emissions. Embodied carbon is the sum of the impact of all carbon emissions linked to the material life cycle. Currently, carbon emissions associated with the life cycles of materials for buildings comprise approximately 11%–12% of all global carbon emissions, including linked energy- and process-related carbon emissions. The main factors of building-related embodied carbon include construction technique, material demand, durability, origin (recycled or reused versus virgin and location), composition of materials, manufacturing processes, and the ability to reuse or recycle. Globally, cement and steel are two of the most significant sources of building-material-related carbon emissions. Total cement production is responsible for approximately 7% of global carbon emissions, while steel contributes 7%–9% of the global total, of which approximately half can be connected to buildings and construction.

Reaching net-zero embodied carbon for major building materials, such as cement and steel, is highly challenging, as these sectors are among the most difficult to decarbonise. Coordinated actions are required to decrease the demand for materials and

DOI: 10.1201/9781003177708-7

promote changes to low-carbon materials, maximise energy efficiency in material production, and transition from carbon-intensive sources of energy to renewable sources. From a decarbonising perspective, there are also opportunities to develop systems to enable a circular economy, reuse, and recycle construction materials. Reaching this target of decarbonisation involves engaging all stakeholders along the value chain, obtaining clear information and robust data on embodied carbon concerning different materials, implementing management systems in industry, revising building standards and codes, using building certification systems, green public procurement, and virgin material taxation. In the long run, codes and regulations on the limits of embodied carbon levels must be developed and applied to ensure that embodied carbon is considered during policymaking and planning and to decarbonise the energy system. Furthermore, it is important to assess the embodied carbon and to establish performance targets for reduction over the baseline over time. Specific targets should be set for material efficiency, energy efficiency of production, and decarbonisation technologies in all subsectors, particularly for the major materials used, such as cement and steel, while promoting low-carbon and nature-based solutions for building materials.

Lifecycle analysis (LCA) quantifies the environmental impacts of material extraction through product manufacturing, usage, reuse, repair, remanufacturing, recycling, and end of life. Decisions regarding building usage, design, and choice of materials should consider the entire lifetime of the building and its materials and products. National, regional, or international databases containing information on the embodied energy and carbon of construction materials are necessary to implement LCA to select proper design approaches. The design of a construction project should focus on minimising carbon emissions using LCA estimations. Thus, all stages of the construction project process should be considered and planned. Considerations of how residuals and waste will be managed by providing options to recover, reuse, or recycle materials should be established as early as possible in construction projects.

Lifecycle stages and modules for a building's life cycle can be described according to the European standard EN15978 and Figure 7.1. The building lifecycle stage model described below is often used when targeting carbon emission reduction roadmaps regionally and by single countries and is divided into four main stages:

- *Product stage (A1–A3)* – Extraction of raw material (A1), transport to the production facility (A2), and manufacturing of product from the raw materials (A3)
- *Construction process stage (A4–A5)* – Transport from the manufacturer to the construction site (A4) and the construction or installation work processes (A5)
- *The use stage (B1–B7)* – The actual use (B1), maintenance work (B2), repair (B3), replacement of materials or equipment (B4), refurbishment such as renovating or redecorating (B5), operational energy use (B6), and operational water use (B7)
- *End-of-life stage (C1–C4)* – Work concerning deconstruction and/or demolition (C1); transport of deconstructed material and products (C2); waste processing such as treatment of the deconstructed materials and products for reuse, remanufacture, or recycling (C3); and disposal of unwanted, hazardous, or dangerous components out of the circular loops (C4)

Building Life Cycle Information																			Supplymentary information
A 1–3 Product stage			A 4–5 Construction process stage		B 1–7 Use stage								C 1–4 End of life stage				Supplementary environmental info
A1 – Raw material supply	A2 – Transport	A3 – Manufacturing	A4 – Transport	A5 – Construction-installation process	B1 – Use	B2 – Maintenance	B3 – Repair	B4 – Replacement	B5 – Refurbishment	B6 – Operational energy use	B7 – Operational water use	C1 – De-construction, demolition	C2 – Transport	C3 – Waste processing	C4 – Disposal	Biogenic carbon storage	Net exports of locally produced electricity

Figure 7.1 Lifecycle stages and modules for a building's lifecycle according EN 15978 (Boverket (2022)).

Supplementary environmental information on carbon emission treatment offsite to the building's environment is also available as a supplement to the above stages, including biogenic carbon storage and net export of locally produced energy.

The strategy for reducing embodied carbon is to establish overall targets for construction projects. This must rely on adequate data collection and the development or adaptation of standardised tools. It is necessary to create and expand databases for generic (average) data of embodied carbon for material categories and to create and expand the availability of the Environmental Product Declaration (EPD) regarding specific building products and materials. An EDP, a standardised document based on the international standard ISO 14025 type III declaration, contains precise data about a specific material's environmental impact, including carbon emissions (see an example of an EPD in Figure 7.2).

An EPD could contain (per the example in Figure 7.2) information on the manufacturer, the manufacturing process, descriptions of the actual product or sample of products, including information on the product or product mass density (important input data when calculating carbon emissions), how the LCA was assessed, what environmental indicators were used, the performance of the indicators including the embodied global warming potential (GWP), GWP in kg CO_{2ekv} (where CO_{2ekv} is all GWP emissions converted to equivalents of carbon emissions), additional information concerning environmental issues, tracking of differences from previous versions of the EPD, and references of methods used to achieve the EPD result.

Figure 7.2 Example of an EPD (Environdec (2022)).

A standard procedure of estimating embodied carbon emission

First, the basic conditions for estimating the embodied carbon emissions of buildings must be established. A summary of the amount of the specific material required to construct the building must be compiled, the weight of the materials must be calculated, and energy and fuel use must be tabulated. This is usually referred to as a bill of resources (BOR). Ideally, the BOR should be completed in conjunction with the project cost estimation. In the early stages of a project, when it is not yet known what products or materials will be included, approximate key indicators and experience-based values are, sometimes, used to estimate the amount of material needed. However, an approximate cost estimation is often created in the early stages to provide a rough summary of the resources required for the project. Once the BOR (energy, materials, and products specified in kg or kWh) is completed, the resources are then linked to various climate data, and the climate impact of the building is then estimated. If the cost estimation software does not automatically convert material amounts to climate impact, it should be completed by using a different tool to convert volumes of construction products and materials to generic resources, e.g. average values of carbon emissions, or to specific resources by EPDs. Generic data should be representative of intended construction products for the local market of construction materials and are often set in a conservative value, which implies a significantly higher value than the true average value. At a later stage of the construction process, when it is better known what products and materials will be used, product-specific data and EPDs can be substituted for generic data. If product-specific data are not available, generic data can be used for the final estimation at the end of the construction process.

The basic stages of a building's life cycle must also be established, preferably using the lifecycle stages mentioned above (see Figure 7.1), which is a useful structure of a building's life cycle. These stages are synchronised with tools for estimating the carbon emissions of a construction project.

The next stage is to obtain data on the material's specific climate impact during the stages of the building's life cycle (see above) in kg CO_{2ekv}/kg. One option is to obtain generic data from international, national, regional, or commercial databases, in which groups of similar material categories are collected with a weighted average value of climate impact. Available tools for carbon emission estimations are mostly connected to databases of generic data. Another option is to obtain data from a specific manufacturer of the material in question using its EPD, either directly from the manufacturer or from special free EPD databases (international or national). The weight of the material is then multiplied with the carbon emission data, generic or specific as above, for every lifecycle stage in question, and the total amount of carbon emissions in kg CO_{2ekv} is then summarised. The above calculation has to be repeated for every material in question, and lastly, a total sum of all actual material carbon emissions must be calculated. Finally, a relation of this amount of carbon emission must be applied to the actual building. One solution is to relate to the building's total net gross area, such as kg CO_{2ekv}/m$^2_{\text{net gross area}}$.

Case 3: How to reduce embodied carbon in practice – an example from Sweden

From 2022 onward, it is mandatory in Sweden to create a carbon emission declaration for every new building that comprises more than 100 m^2 floorspace. The declaration is meant to be a report to improve knowledge of and reduce the climate impact of the actual building, and to reach the target of zero carbon by 2045. The Act on Climate Declarations for Buildings under Construction was implemented in Sweden on the 1st of January 2022, along with appropriate ordinances and provisions. The legislation is to be harmonised with the European Commission's Level(s) framework (see Section 7.4.2 in this chapter), and the European EN 15978:2011 standard for the environmental performance of buildings.

According to the Act, a climate declaration describes the building's climate impact, based on the calculated carbon emissions from the construction stage. This includes the A1–A3 and A4–A5 stages previously described. The purpose of this regulation is to reduce the climate impact of new buildings and to increase knowledge of the climate impact of building construction. Another purpose is to illustrate the benefits of reducing the climate impact of building construction to various stakeholders in the construction sector. A climate declaration provides the client with more in-depth knowledge of resource flows in the building, and its carbon emission impacts and documented climate performance. Climate estimation in the early stages of a construction project process provides a quantitative basis for making well-grounded decisions on how a building's climate impact can be reduced. Increased knowledge also provides opportunities to reduce the amount of material and construction residuals. In addition, the costs of the construction project could decrease in both the short and long term.

It is the client's responsibility to produce a climate declaration and register it in the national database. This responsibility is connected to the building permit, which is not granted final clearance without a registered climate declaration. The climate declaration requirement applies to all new buildings that require a building permit, with some exceptions:

- Buildings with a temporary building permit, intended to be used no more than two years
- Buildings for industrial facilities and workshops, agricultural buildings for farming, forestry, and other similar enterprises
- Buildings that do not have a gross floor area larger than 100 square meters
- Buildings intended for defence purposes and buildings of importance to Swedish security
- A client who is a natural person and does not construct a building in business does not need a climate declaration
- Some buildings for which the client is public

The climate declaration is registered with the Swedish National Board of Housing, Building, and Planning (Boverket), which has a register and database for received climate declarations. The declaration can be registered digitally by the client. Gathering information for the construction project as early as possible in the process by comparing different solutions of design and material choices creates a better chance for taking measures for climate improvements. Thus, saving and gathering information throughout the entire construction process also simplifies work associated with the final climate declaration because it is more time-consuming to gather information afterwards.

The content of the climate declaration

The Swedish regulated climate declaration covers the entire construction stage, including the following five modules in the construction stage (A1–A5), as described previously and according to the European standard EN15978:2011, during the life cycle. This implies that climate calculations can be presented in a uniform manner, which facilitates interpretation of the calculated results. Currently, the entire life cycle does not include the climate declaration, such as the use stage B1–B7 that considers climate impact from operational energy and the end-of-life stage, C1–C4 during deconstruction. Nevertheless, the Swedish government plans to increase the scope of the declaration as a continuing roadmap until 2045. It is, of course, voluntary to estimate the building's entire life cycle with the 2022 version of declaration requirements; only data for modules A1–A5 must be submitted in the climate declaration implemented at Boverket in 2022. Concerning module A4, only the transport of construction products to the construction site is included in the declaration, and for module A5, only construction residuals and energy-demanding processes are included in the declaration.

The 2022 version of the Climate Declaration is limited to certain building elements and components. The building envelope, load-bearing structures, and interior walls are included. This implies, for example, that the HVAC installation and ventilation equipment of the building are excluded. In the context of declaration requirements, the building envelope consists of one or more layers that insulate the inside of the building from its surroundings in terms of temperature, noise, and moisture, among others. Load-bearing structures are part of a building's construction that bear loads of various types in addition to their own weight. Concerning interior walls, interior surface layers on the building envelope, interior surface layers on load-bearing structures, and surface layers on interior walls are excluded as requirements of the climate declaration.

Generic climate data are available in a database developed by Boverket on its website (https://www.boverket.se/en/start/building-in-sweden/developer/rfq-documentation/climate-declaration/climate-database/). The generic climate data in the database for construction products and materials are set conservatively, i.e. with a relatively high above average value (approximately 25%), to stimulate the use of EPDs in the final climate declaration. When using generic climate data, it must be derived from the database of Boverket and no other. Different LCA-based tools that facilitate climate calculations and form the basis for climate declaration are available in the market.

Limit values

Limit values could have been introduced at the end of the 2020 decade based on the proposed roadmap, but decisions by the government have not yet been made. Concerning the first level of conditions, enacted in 2022, the declaration should contain carbon emission impacts from load-bearing structures, the building envelope, and interior walls, all covering lifecycle stages A1–A5. No limit has been determined at this stage for the amount of carbon emissions from the building. The next level, proposed by 2027, should establish a limit of 20% below an average value during stages A1–A5 from incoming declarations of the national register between 2022 and 2026. The building parts are complemented by installations, interior surface finishes and fitments, and a more complete representation of a building. The lifecycle stages are complemented by the use stage (B2, B4, and B6) and the deconstruction stage (C1–C4), as shown in Figure 7.1. In addition, impacts from biogenic carbon storage in wood-based materials and products and net exports of locally produced electricity should also be declared. For the use stage, the reference period for the calculation was set to 50 years. From 2035, the limit is set to 40% below the reference value; from 2045, the limit is set to 80% below.

Unsubmitted climate declaration

Without the climate declaration, the municipality cannot issue any final clearance for the building in question according to the Swedish Planning and Building Act. In such cases, the municipality can only issue an interim final clearance. When the client registers the climate declaration, a receipt from Boverket is given; thereafter, the client must present this receipt to the municipality to obtain the final clearance.

After a climate declaration is submitted, it is placed in a climate declaration register among all other submitted declarations for which Boverket is the leading responsible authority. There is no statement indicating whether the declaration is 'approved' or 'rejected'. A requirement is that the client must save basic information for the climate declaration, such as the bill of resources and eventually used EPDs, for at least five years. As the main authority, Boverket can request this basic information from the client; if a control by calculated climate impact value significantly deviates from the registered declared value, the client has the opportunity to submit an explanation. If the client does not submit a reasonable explanation for the deviation, the controlled value can be registered, and the client may also be charged a sanction fee.

7.2 General assessment methods

Certification and labelling are useful when they provide an objective evaluation of the environmental impact of a material, product, or process. There are schemes available to audit several construction-related products, including paints, fabrics, cleaning products, materials, white goods, and buildings. The latter includes energy efficiency, noise, water consumption, biodiversity, or air quality. In addition, labels are used for construction processes, larger projects, and professional practices. Such labels enable clients, contractors, suppliers, and buyers to express a preference for desired requirements regarding buildings, materials, and products or to obtain impartial confidence in the integrity of a professional. Labels are shared with charities, industry and trade bodies, governments, and other interested parties. Several different general assessment methods can be used to audit the various impacts of activities on the environment, energy, and sustainability, including:

- *Ecological footprint* – A tool enabling an estimation of resource consumption and assimilation of waste of a defined human-populated area or economy in a corresponding productive land and water area. The ecological footprint of a population, economy, individual building, or person can be estimated as the relative measured value. The national footprint calculated as national production + imports – exports provides basic data for all ecological footprints worldwide. Thus, a country has an ecological reserve and an ecological credit if its footprint is smaller than its biocapacity area. Otherwise, it is in ecological deficit as an ecological debtor. Most countries, and the world as a whole, are experiencing ecological deficits. The entire world's ecological deficit is referred to as global ecological overshoot, and as Earth Overshoot Day, it is manifested when, during a single year, humanity's demand for ecological resources and services exceeds what Earth can regenerate in that year.
- *MIPS* – material input per unit of service, for example, all inputs of materials influenced by human use (ores, minerals, soil, etc.), differentiated by categories and calculated in kilograms. The method encompasses all materials to be moved to take advantage of raw materials or to build infrastructure, including residual production (overburden, gangue, etc.), drainage water, and logged trees.
- *Ecological rucksack* – The total quantity (kg) of natural materials necessary to produce, use, recycle, and dispose of a product or service minus the actual weight of the product. That is, ecological rucksacks consider hidden material flow, adopting a lifecycle approach and signifying the environmental strain or resource efficiency of a product or service. It is a variant of MIPS (above).
- *Factor X* – The consumption of a resource should not exceed its regeneration and recycling rate or the rate at which all functions can be substituted; the long-term release of substances should not exceed the tolerance limit of environmental biodiversity and their capacity for assimilation; and hazards and unreasonable risks for humankind and the environment due to anthropogenic influences must be avoided. The timescale of anthropogenic interference with the environment must be in a balanced relationship with the response time needed by the environment to stabilise itself. It is also known as factor 10, which is an estimated general requirement. X represents the diversity of requirements depending on the country's content in different contexts.
- *Energy assessments* – Energy performance certificates suitable for electronic equipment, white goods, building equipment, or whole buildings (see Section 1.6 in Chapter1).
- *Different building equipment schemes* – Ventilation operation assessment where indoor air quality, filter exchanging routines, energy performance, and availability of instruction for use and maintenance are controlled.
- *Environmental impact assessment* – Required when developing or refining an area for a new purpose of use, mandatory in some countries by law. In reality, it is assessed only when it concerns major industry-development projects, power plants, power infrastructure, major wind power facilities, roads, or rail projects.

Other tools exist for the assessment of materials, resources, products, places, occupant satisfaction, components, buildings, professions, processes, projects, leadership, social factors, business performance, and investment, and sometimes aspects of these are merged into a joint assessment. These issues often include energy and water use, light quantity, emissions, land use, cost, speed, design quality, noise, waste, volatile organic compounds, embodied energy, embodied toxicity, quality of place, quality of life, professional competence, indoor air quality, biodiversity, transport, occupant satisfaction, and business impact.

In addition to pure environmental gains, the benefits of using such assessment tools regarding the built environment include creating a base for ensuring the existing and forthcoming value of the actual property and providing validation to ensure lower insurance costs of the property. It is also beneficial regarding implementation of the client's environmental management system, such as ISO14001, and to ensure and verify such continual improvements. Further, it provides a base to negotiate better terms of financing, determine property taxation and property transfer, and promote user satisfaction and gain better market opportunities for the client.

7.3 Selection of environmentally labelled construction materials and products

Labels and certifications regarding materials and products are useful for clients and design teams when recognising that actual regulations are inadequate to provide resilience against future challenges and when developing more resilient solutions. They often include a combination of requirements and encompass elements of peer reviews. The monitored and measured performances are usually set against traceable quantitative performance standards. Labels or certifications can be particularly useful in motivating design teams to increase their knowledge of sustainability when designing and meeting client objectives. However, labels and certifications are rarely sensitive to context and mostly reward quantifiable elements, and rarely qualitative elements. One important detail when selecting construction materials is knowledge of chemical reactions between two or more seemingly harmless materials when they are incorporated into one functional entity in a specific environmental context.

Commercial databases of environmentally assessed materials and products are available. Some of the most common international schemes for materials include the following.

- *The Forestry Stewardship Council (FSC)* – A global certification that confirms forestry management that preserves biological diversity beneficial to local people and workers while ensuring the economic viability of raw material for timber and wood products
- *PEFC (Programme for the Endorsement of Forest Certification) scheme* – Similar to the FSC above
- *Cradle2Cradle* – A global standard for products that are safe, circular, and responsible. There are requirements for continually improving which products and how they are made into five quality categories: material health, product circularity, clean air and climate protection, water and soil stewardship, and social fairness
- *The EU Ecolabel* – Materials assessed by the European Union official label as a label of excellence concerning products and services that meet high environmental

standards throughout the life cycle, from raw material extraction to production, distribution, and disposal

- *The Fairtrade Marks* – A global ethical scheme for how a product is manufactured with respect to human rights and a good working environment for employees, which should be fully traceable

Various established regional and national eco-labels include the following:

- *UKWAS (UK Woodland Assurance Standard)* – Certification for verifying sustainable woodland management in the United Kingdom used for both FSC and PEFC certification
- *The Green Guide to Specifications* – United Kingdom-based guidance on how to make best environmental choices when selecting construction materials and products
- *The European Energy Label* – White goods are labelled according to their energy efficiency, from A+++ to G
- *Natureplus eco-label* – A German-based label specialising in building materials, products, and furnishings. It certifies only products containing a minimum of 85% recycled materials or that are mineral-based. Further requirements are risk to human health and consideration of lifecycle analysis.
- *The Nordic Swan ecolabel* – Developed primarily for the Scandinavian market in Denmark, Sweden, and Norway. The label includes information on a wide range of products, including many relevant to the construction industry, and also details the criteria that each product must meet to gain certification. In addition, it contains criteria to certify whole buildings, such as small houses, apartment buildings, and buildings for schools and pre-schools.
- *BASTA* – Building materials assessment criteria in Sweden, assessed by impartial assessors in three categories concerning the chemical content of dangerous or hazardous emissions, use of toxic constituents during production of the material, and so on.

Larger contractors and real estate enterprises have created lists of materials and products to avoid, such as those containing specific dangerous chemicals. When uncertainty prevails, the precautionary and substitution principles should be used as guidelines. If there are requirements for verifying materials or other criteria from building environmental certifications during design or construction work, knowledge of how to execute and verify these certification criteria is of great importance.

7.4 Building rating and certification schemes

Various building ratings and certification schemes exist globally in relation to sustainability, energy rating, health, and well-being as well as some related to regenerative building certifications. Below are most of the rating systems listed by the World Green Building Council as well as various rating systems from other sources. As some could be missing and some could be out of date, it is not a complete list of certification schemes; however, it offers an example of the global diversity of existing schemes. (When referenced in the list, GBC represents the local or responsible regional Green Building Council in question.) The following list of building certification schemes appears in alphabetical order without priority regarding sustainability or regenerative content and includes relevant Internet links for each, some with slightly more details of the scheme:

- ARZ Building Rating System (Lebanon GBC), https://lebanon-gbc.org/arz-building-rating-system/
  - o To measure existing commercial buildings in Lebanon, the extent of healthy, comfortable places for working, consuming the right amount of energy and water while having a low impact on the natural environment
- Assessment Standard for Green Building of China (China Academy of Building Research), https://www.worldgbc.org/sites/default/files/Introduction%20to%20China%20Green%20Building%20Assessment%20Standard%203rd%20Edition.pdf
- BeamPlus (Hong Kong), https://www.hkgbc.org.hk/eng/beam-plus/introduction/index.jsp.
  - o A set of performance criteria for a range of sustainability issues relating to the planning, design, construction, commissioning, fitting-out, management, operation, and maintenance of a building. Provides an objective assessment of a building's overall performance throughout its life cycle.
- BERDE (Philippine GBC), http://berdeonline.org/#about-berde
  - o Building for ecologically responsive design excellence is a tool for assessing, measuring, monitoring, and certifying the performance of green building projects above and beyond existing national and local building and environmental laws, regulations, and mandatory standards.
- B.E.S.T (Cebdik Turkish GBC), https://cedbik.org/en
  - o The ecological and sustainable design in buildings residential certification system and commercial certification system can be implemented in new residential and commercial projects in Turkey. The evaluation is based on nine categories.
- BREEAM (Building Research Establishment [BRE], UK), https://www.breeam.com/?cn-reloaded=1
  - o The Building Research Establishment's Environmental Assessment Method originated in the United Kingdom and has been locally adapted in numerous countries. A more detailed description is presented in the next section.
- Casa Colombia (Consejo Colombiano de Constrccion Sostenible, CCCS), https://casa.cccs.org.co
  - o A certifying scheme for sustainable and healthy housing projects in Colombia
- CASBEE (Japan Sustainable Building Consortium [JSBC]), https://www.ibec.or.jp/CASBEE/english/
  - o The comprehensive assessment system for building environmental efficiency is a method for evaluating and rating the environmental performance of buildings and the built environment in Japan. It has been designed to both enhance the quality of people's lives and reduce the lifecycle resource use and environmental impacts associated with the built environment, from a single home to an entire city.
- CEEQUAL (Building Research Establishment [BRE], UK), https://www.bregroup.com/products/ceequal/
  - o Civil engineering environmental quality assessment and award scheme for guiding, monitoring, and assessing the quality of infrastructure and civil engineering projects.

- Citylab (Sweden GBC), https://www.sgbc.se/certifiering/citylab/ (Swedish only)
  o A four-part Swedish certifying scheme regarding global and national sustainability objectives for city development. The certification parts are divided into sustainability programmes, action plans for city development, action plans for subprojects, and urban district sustainability.
- DGNB System (German Sustainable Building Council), https://www.dgnb-system.de/en/system/
  o Deutsche Gesellschaft für Nachhaltiges Bauen (DGNB), the international version, is highly flexible regarding country-specific conditions with a lifecycle-based approach. A more detailed description is presented in the next section.
- DGBCWoonwerk (The Dutch GBC, The Netherlands), https://www.dgbcwoonmerk.nl
  o For single Dutch homes, a scheme of objective targets energy consumption, material use, health, flexibility, and comfort but also other important themes, such as safety, accessibility, and living environment.
- E$^+$ C$^-$ Énergie Positve & Reduction Carbone (Executive Council for Construction and Energy Efficiency, CSCEE, France), http://www.batiment-energiecarbone.fr/en/how-the-trial-scheme-works-a3.html
  o The scheme includes calculations of indicators relating to a building's energy consumption and environmental performance, notably, greenhouse gas emissions. It is suitable for France, which is aimed at construction operations involving new buildings from the time the building received final approval. The scheme could be applied during other phases of the construction process, but updated calculations must be performed once the building is given final approval.
- EDGE (International Finance Group [IFC], a member of the World Bank Group, UK, and international), https://www.edgebuildings.com
  o Excellence in design for greater efficiencies is an internationally measurable solution to prove the business case for building green and unlocking financial investment. It includes a cloud-based platform to calculate the cost of going green and utility savings using a set of city-based climate and cost data, consumption patterns, and algorithms for predicting the most accurate performance results. Available for zero carbon certificate.
- Estidama Pearl Building Rating System (Abu Dhabi Urban Planning Council [UPC], UAE), https://pages.dmt.gov.ae/en/Urban-Planning/Pearl-Building-Rating-System
  o The aim of the Pearl Building Rating System (PBRS) is to promote the development of sustainable buildings and improve quality of life. Achieving sustainability for buildings is a requirement to integrate the Estidama (Arabic for sustainability), aimed at transforming Arabic cities into a model of sustainable urbanisation. The PBRS encourages water, energy, and waste minimisation and local material usage, and improves supply chains for sustainable and recycled materials and products.
- GBC Brasil Casa & Condominio (Brazil GBC), https://www.gbcbrasil.org.br/certificacao/certificacao-casa/
  o A certifying scheme for sustainable and healthy housing projects in Brazil
- Greenship (GBC Indonesia), https://gbcindonesia.org/certification
  o Greenship comprises six types of certification schemes for Indonesia: new buildings, existing buildings, interior spaces, homes, neighbourhoods, and net zero health.

- Green Building Index (GBI Malaysia), https://www.greenbuildingindex.org
  - o The focus is on increasing the efficiency of resource use, energy, water, and materials while reducing the impact of buildings on human health and the environment during the building's life cycle. This should be accomplished through improved siting, design, construction, operation, maintenance, and removal. Green building certification for Indonesia should be designed and implemented to reduce the overall impact of the built environment on its surroundings.
- Green Key Global (Canada), http://www.greenkeyglobal.com
  - o A global scheme developed in Canada to certify hotels, motels, and resort commitment to improving environmental and fiscal performance.
- GREENSL (GBC Sri Lanka), https://srilankagbc.org/greensl-rating-system/
  - o The rating system for existing and new buildings in Sri Lanka is a performance standard used to certify the operations and maintenance of commercial or institutional buildings of all sizes, both public and private. The intent is to promote high-performance, healthy, durable, and affordable environmentally sound practices in existing buildings. In addition, it aims to implement sustainable practices and reduce the negative environmental impacts of buildings on the functional lifetime of clients and operators.
- Green Star (Australia and New Zealand GBC), https://new.gbca.org.au/green-star/exploring-green-star/
  - o The Green Star is an internationally recognised rating system for healthy, resilient, and positive buildings and places. It includes reducing the impact of climate change; enhancing health and quality of life; restoring and protecting the planet's biodiversity and ecosystems; driving resiliency in buildings, fitouts, and communities; and contributing to market transformation and a sustainable economy.
- HomeStar (New Zealand GBC), https://www.nzgbc.org.nz/homestar
  - o HomeStar is an independent rating tool for assessing the health, efficiency, and sustainability of homes across New Zealand by rating a home's performance and environmental impact.
- GBC Home (GBC Italy), https://www.gbcitalia.org/web/guest/home1
  - o GBC Home is an Italian housing rating system based on the Leadership in Energy and Environmental Design (LEED) and Italian context. It promotes health, durability, economy, and best environmental practices in the design and construction of buildings.
- GBCSA Net Zero/Net Positive (South Africa GBC), https://gbcsa.org.za/certify/green-star-sa/net-zero/
  - o Net Zero/Net Positive Certification refers to projects that go beyond the partial reductions required in current GBCSA schemes (see below) and aim to reach the endpoint of completely neutralising or positively redressing their impacts.
- Green Star SA (South Africa GBC), https://gbcsa.org.za/certify/green-star-sa/
  - o Special versions of rating schemes suited to South Africa and other African countries. The schemes provide objective measurements for green buildings in South Africa and other Africa-based countries on nine different categories, each with a range of credits that address the environmental and sustainability aspects of designing, constructing, and operating a building. Locally adapted to Botswana, Tanzania, Uganda, Ghana, Rwanda, Kenya, Nigeria, Mauritius, Namibia, and Morocco.

- GRESB (The Netherlands and International), https://gresb.com/nl-en/
  - The Global Environmental, Social, and Governance Benchmark for Real Assets is a mission-driven and investor-led organisation providing actionable and transparent data by benchmarking real assets globally of environmental, social, and governance performance to financial markets with the aim of creating a sustainable world.
- GSAS Certifications (GSAS Trust, Qatar, Middle East, and North Africa [MENA region]), https://gsas.gord.qa
  - Global Sustainability Assessment System, GSAS, certification schemes aim to create a sustainable built environment that minimises the ecological impact while addressing specific social and cultural needs and the environment of the MENA region. There are three types of certification schemes: design and build for new buildings and infrastructure projects, construction management for ongoing construction and infrastructure projects, and operations for existing buildings and infrastructure projects.
- Home Performance Index (GBC Ireland), https://homeperformanceindex.ie
  - The home performance index is Ireland's national certification for new homes. Similar to the certification for commercial development of LEED and BREEAM (see next section), except that it is specifically designed for residential development and connects to Irish building regulations, EU CEN standards, and international WELL certification (see the next section) for communities.
- HQE (GBC France and International, CERQUAL, and CERTIVEA), https://www.behqe.com/home
  - High environmental quality is a French and international scheme to obtain objective and credible solutions to certify best practices in terms of the sustainable construction and operation of buildings and high-quality and sustainable local developments. Certification encompasses the entire life cycle of a building, such as construction, renovation, and operation, and includes non-residential buildings, residential buildings, detached houses, and urban planning and development. Internationally, the scheme has been locally adopted for Lebanon, Luxembourg, Morocco, Spain, Brazil, and Québec, Canada.
- IGBC Green Building Rating systems (India GBC), https://igbc.in/igbc/
  - Rating systems include sustainable practices and solutions to reduce environmental impact. Green building design provides an integrated approach that considers the lifecycle impact of the resources used. The 31 IGBC rating systems are voluntary, consensus-based, and market-driven building schemes based on five elements of nature (Panchabhutas), and are a blend of ancient architectural practices and modern technological innovations. The rating systems were applicable to all five climatic zones of India.
- INSIDE/INSIDE (Dutch GBC, The Netherlands), https://www.insideinside.nl/en
  - INSIDE/INSIDE is an independent scheme for a sustainable interior, based on an online platform where the sustainability of products and materials is compared in an objective and presentable way. The scheme aims to make the interior sector in the Netherlands more sustainable by sharing the latest knowledge and information on sustainability, circularity, and connecting partners.
- LOTUS (GBC Vietnam), https://vgbc.vn/en/rating-systems/
  - LOTUS is a set of voluntary green building objective rating systems developed in established building physics science and adapted to the conditions, climate, and regulations of the Vietnamese construction sector. The scheme provides

a holistic assessment of the environmental performance over the life cycle of buildings with an integrated approach for evaluating buildings in terms of energy consumption, water use, waste management, and indoor environmental quality. It can be used for decision-making by stakeholders, developers, clients, designers, and contractors.

- LEED (US GBC and International), https://www.usgbc.org/leed
  - o US-based LEED, which measures the sustainability of building construction projects, has also been adopted in Canada, India, and several other countries. A more detailed description is presented in the next section.
- Living Building Challenge ([LBC] International Living Future Institute, US and International), https://living-future.org/lbc/
  - o The LBC scheme raises the bar above widespread sustainability rating and certification tools. Its main objective is to eliminate any negative impact that a building might have on the global environment and health and to promote positive regenerative development. A more detailed description is presented in the next section.
- Level(s) (European Commission, EU), https://ec.europa.eu/environment/levels_en
  - o This European framework for sustainable buildings provides a common language for assessing and reporting the sustainability performance of buildings, offering an entry point for applying circular economic principles to the built environment. It uses sustainability indicators to measure the carbon, materials, water, health, comfort, and climate change impacts throughout a building's full life cycle. At first glance, Level(s) could be mistaken as a building certification scheme, but unlike such schemes, it comprises more of a set of tools to help stakeholders regarding many aspects of interpretations of sustainability. More details are provided in the section on critical path.
- Miljöbyggnad (Sweden GBC), https://www.sgbc.se/app/uploads/2021/12/210701-MB-A4-ENG-Utfall.pdf
  - o A nationally developed certification scheme for Sweden adapted to national codes and regulations concerning buildings, construction, and the built environment.
- NABERSNZ (New Zealand GBC), https://www.nabersnz.govt.nz
  - o The NABERSNZ is a scheme for measuring and improving the energy performance of office buildings in New Zealand.
- NollCO$_2$ (Zero carbon) (Sweden GBC), https://www.sgbc.se/certifiering/nollco2/
  - o NollCO$_2$ is a nationally adapted Swedish scheme for reaching zero carbon emissions. It is an addition to other Swedish-adapted certification schemes, such as Miljöbyggnad, BREEAM-S, and LEED, and their highest rating level.
- Nordic Swan Ecolabel (Nordic Ecolabel), https://www.nordic-ecolabel.org/product-groups/group/?productGroupCode=089
  - o A scheme for multi-store buildings, schools, daycare institutions and houses, detached houses, terraced houses, and holiday homes with requirements throughout the life cycle. Suitable and adapted for Nordic countries, including Denmark, Norway, Sweden, Finland, and Iceland.
- One Planet Living (Bioregional, UK and International), https://www.bioregional.com/resources/one-planet-living-for-sustainable-places
  - o Provides a framework for creating more sustainable places by publishing an action plan based on ten planet living principles; no certification is needed. Suitable for the United Kingdom and worldwide.

- Parksmart (Parking Facilities US and International), https://parksmart.gbci.org/certification
  - The Parksmart scheme comprises the only rating system for advancing sustainable mobility through smarter parking structure design and operation.
- Passive House (Passive House Institute, Germany and International), https://passivehouse.com
  - The Passive House scheme provides a building standard that is intended to be energy efficient, comfortable, and affordable, with a concept based on reducing heat losses to an absolute minimum.
- PEER (US Green Business Certification Inc., GBCI), https://peer.gbci.org
  - Performance Excellence in Electricity Renewal (PEER) is a certification programme that measures and improves power system performance and electricity infrastructure.
- SEED, Pakistan Green Building Council Guidelines (Pakistan GBC), https://pakistangbc.org/pgbc-guidelines.php
  - Sustainability in Energy & Environmental Development (SEED) is a Pakistani rating system for new and all types of buildings in all phases. It concerns sustainable and socially responsible thinking about the planning, construction, maintenance, and operation of buildings and communities.
- SGBC Green Certification (Singapore GBC), https://www.sgbc.sg/sgbc-certifications
  - The two certification schemes for the environmental standards of building products and services that provide benchmarks for environmental performance include the Singapore Green Building Product (SGBP) certification scheme and Singapore Green Building Services (SGBS) certification.
- SITES (US Green Business Certification Inc., GBCI), https://www.sustainablesites.org
  - SITES is a rating system designed to distinguish between sustainable landscapes, measure their performance, and elevate their value. The scheme is aimed at development projects located on sites with or without buildings, ranging from national parks to corporate campuses, streetscapes, homes, etc.
- Swiss DGNB System (SGNI, Switzerland), https://www.sgni.ch (German) and http://www.cuepe.ch/html/enseigne/pdf/trp-15-16-08.pdf (English).
  - The Swiss version of DGNB
- TRUE (US Green Business Certification Inc., GBCI), https://true.gbci.org
  - TRUE comprises a whole systems approach aimed at changing how materials flow through society, resulting in no waste. This encourages the redesign of the resource life cycles of the products to be reused. The US scheme promotes the processes of the entire life cycle of the products used within a facility.
- VERDE (GBC Spain), https://gbce.es/certificacion-verde/
  - A five-part Spanish certification scheme for sustainable buildings
- The WELL Building Standard (The International WELL Building Institute, US and International), https://www.wellcertified.com
  - WELL certification defines the requirements for healthier buildings that improve users' well-being and productivity in 11 areas. A more detailed description is presented in the next section.

More of the most common certification schemes

Some of the most widely used rating schemes or tools for construction projects globally are LEED, BREEAM, DGNB, WELL, and LBC. The following is a brief description of these schemes.

- *BREEAM (Building Research Establishment Environmental Assessment Method)* – First published in the United Kingdom in 1990, the BREEAM comprised the world's first sustainability assessment method for master plans, infrastructure, and buildings. It recognises and reflects the value of higher-performing assets across the built environment life cycle, from new construction to use and refurbishment. Since then, it has been applied in more than 90 countries. BREEAM contains requirements for the building scale in the following categories: management, health and well-being, energy, transport, water, materials, waste, land use and ecology, pollution, and innovation. A BREEAM-certified rating reflects the performance achieved by a construction project and its stakeholders, as measured against the scheme and its benchmarks. The rating is divided into different levels from Acceptable (In-Use scheme only) to Pass, Good, Very Good, Excellent, and Outstanding, and is reflected in a series of stars on the BREEAM certificate. Each category is subdivided into a range of assessment issues, each with its own aims, targets, and benchmarks. When a target or benchmark is reached, as determined by the BREEAM assessor, the development or function score points are called credit. The category score is then calculated according to the number of weighted credits achieved. Once the development has been fully assessed, the final performance rating is determined by the sum of weighted category scores. The scheme is suitable for different types of buildings, such as residential, office, public, healthcare, retail, and industrial. It also serves as a scheme for refurbishment projects as well as infrastructure and community master planning.
- *LEED* – Developed in the United States by the US Green Building Council, LEED is a voluntary and market-driven rating tool that measures sustainability for construction projects. Since the launch of its first version in 1998, it has become one of the most internationally widespread construction sustainability assessment tools. LEED contains mandatory and optional requirements for nine categories: integrative design, sustainable sites, location and transportation, water efficiency, energy and atmosphere, indoor environmental quality, materials and resources, innovations, and regional priorities. To achieve LEED certification, the construction project must collect points based on requirements and credits from the categories above. After a verification and review process, points are earned to meet a level of certification: certified (40–49 points), silver (50–59 points), gold (60–79 points), and platinum (80+ points). The scheme is suitable for different types of buildings, such as residential, office, public, healthcare, retail, and industrial. In addition, it serves as a scheme for existing buildings, refurbishment projects, infrastructure, community master planning, and when to recertify an already certified building. Furthermore, LEED Zero, a zero carbon scheme, recognises net zero carbon emissions from energy, consumption through carbon emissions avoided or offset, and a

source energy use balance of zero over a period of 12 months. Moreover, it recognises a potable water use balance of zero over a period of 12 months, and buildings that achieve a special waste certification, TRUE certification at the platinum level.

- *DGNB (Deutsche Gesellschaft für Nachhaltiges Bauen)* – The German DGNB system is another variant of the widespread sustainability rating tools with a life-cycle-based approach. An international version that adheres to international requirements or standards has been available since 2020. In cases where no ISO standards were available, European standards were chosen, or the German DIN standards were adapted internationally. The international version is flexible in terms of country-specific conditions. Depending on each general condition, it may be necessary to adapt the criteria or reference values and the weighting of criteria for tailored solutions in diverse countries. Based on three key objectives – LCA assessment, holistic orientation, and performance orientation – it considers the entire life cycle and evaluates the overall performance. The scheme is applicable to new construction projects as well as to existing buildings, refurbishment projects, and buildings in use to measure different functions with different credit weightings. A certificate can obtain level bronze (total performance index greater than 35%), silver (total performance index greater than 50% and minimum performance index from 35%), gold (total performance index greater than 65% and minimum performance index from 50%), and platinum (total performance index greater than 80% and minimum performance index from 65%). The international scheme is suitable for different types of buildings, such as residential, office, public, healthcare, retail, and logistics buildings. Within the scheme, bonuses are set for minimum climate neutrality and beyond regarding carbon emissions.

- *WELL* – The WELL standard, published since 2014 by the International WELL Building Institute, currently offers WELLv2. WELL certification defines requirements for healthier buildings that improve users' well-being and productivity in 11 categories: air, water, nutrition, light, movement, thermal comfort, sound, materials, mind, community, and innovation. The rating tool includes 24 mandatory preconditions and 98 optional optimisations required to reach different certification levels. The optimisations offer a set of options that can be freely selected and applied and include parts that the project team can focus on to meet certification requirements. The certification levels include bronze, silver, gold, and platinum. WELL projects are divided into two main groups, determined primarily by ownership type: owner-occupied and WELL core. Owner-occupied refers to when a project is primarily occupied by a client. WELL core refers to when a project occupies the space mainly by one or more tenants, and only a small part is occupied by the client. Multifamily residents with more than five dwellings in a single building can use certification. The innovation category includes options for scoring points by reducing carbon emissions, and to be recognised as a project to achieve leading green building certificates.

- *LBC* – The LBC standard of the International Living Future Institute (ILFI) is raising the bar above the widespread sustainability rating and certification tools. The main objective of LBC is to eliminate any negative impact that a building might have on the global environment and health and to promote positive regenerative development. The standard defines 20 challenges (called imperatives, each with the same weight) divided into seven categories, called petals. The petals are place, water, energy, health and happiness, materials, equity, and beauty.

The standard is performance-based, and the guiding principles and performance metrics apply regardless of where in the world the project is located. Based on changes related to site-specific circumstances, the design team chooses the most appropriate design approaches. It also includes 'living transects', which originate from biology and ecology science as transects, a cut, or path through a part of the environment that includes a range of different habitats. In the urban context, a zoning system was created – the new urbanism transect model – using the transect methodology mentioned above. This zoning system is intended to replace conventional separated-use zoning systems that usually encourage car-dependent culture and land-consuming sprawl. Applied to all projects, the Living Transect modifies the requirements for and exceptions to numerous imperatives and is an adaptation of the original transect concept and the new urbanism transect model. It addresses issues inherent to suburban zones through requirements that encourage density, food production, and access as well as human-centred versus car-centred development. Living transects comprise natural habitat preservation, rural zones, village or campus zones, general urban zones, and urban centre zones.

The certification options are, depending on desired outcome, as follows:

- *Living certification* – The highest level of sustainability and regenerative design
- *Petal certification* – For construction projects that focus on one of the petals
- *Core green building certification* – For projects that meet the requirements of ten core imperatives, up to two core imperatives per petal, and to verify the performance for water and energy through a 12-month performance period
- *Zero energy certification* – Focused on achieving net zero energy through onsite production of renewable energy
- *Zero carbon certification* – Requires 100% of the operational energy use associated with a construction project to be compensated by new on- or off-site renewable energy. It also requires a targeted energy efficiency level and reduction in the embodied carbon of the project's primary materials. In addition, 100% of the carbon emission impacts associated with the construction and used materials must be reviewed and compensated.

The LBC scheme is suitable for all building types. More details regarding the certification and rating schemes can be found below in the section on Sustainability and regenerative certifications.

Case 4: Two examples of locally adapted schemes

Locally, a few schemes and checklists are available to use, depending on the location of the construction site and the purpose of the construction project. Two examples provided below represent completely different locations and local conditions. The first example concerns Australia, where climate change impacts are real; the second concerns northern Europe which has very different climate conditions and other priorities of sustainability in comparison with Australia.

In Australia, the city of Melbourne maintains a sustainability checklist and fact sheet that provides guidance on new and renovated buildings and concerns

(Continued)

about climate resilience to meet the imminent extreme heat, drought and water scarcity, sea level rise, extreme storms, and flash flooding. It comprises sustainability checklists and fact sheets together with action plans and checklists during the construction process, i.e. design, construction, and maintenance, concerning:

- *Project life cycle* – Including planning and design through lifecycle costs, competitive tendering process in accordance with Melbourne policies, and assessing outcomes through post-occupancy evaluations, research, and energy generation
- *Energy and greenhouse* – Including minimising ongoing energy use through siting and location, providing low-energy thermal control systems and renewable energy generation systems to offset carbon emissions through landscaping, and minimising energy use during construction
- *Materials and waste management* – Including adapting reuse and recycling of materials in design and used materials during construction, optimising use and performance of materials through appropriate manuals and maintenance schedules
- *Biodiversity* – Including protecting and conserving indigenous landscapes, protecting existing habitats, and establishing new and promoting biodiversity of flora and fauna
- *Heritage* – Including protecting and enhancing indigenous and cultural significance and diversity, and ensuring contractor awareness of heritage-related obligations
- *Onsite environmental quality* – Including natural light, glare, temperature control, air quality, long-distance view, passive control system, sun, shade, and wind protection
- *Water consumption* – Including minimised potable water consumption, greywater use and reuse, rainwater collection and reuse, and blackwater use and reuse
- *Stormwater* – Including protecting and improving stormwater quality, minimising stormwater volumes and velocities, beneficial usage of topography, ensuring drainage by natural materials, and regenerating habitats along water bodies

To increase the resilience of buildings to local climate change impacts, the following action is recommended:

- *Heat* – Including measures to prevent and reduce heat in buildings, such as external shading, blinds, high-performance sun radiation glazing, thermal mass usage, appropriate ventilation systems, reduction of internal heat gains, and use of cool roofs
- *Water* – Including measures to reduce water use and flooding risks by water efficient fittings, water recovery end reuse, onsite water storage, flooding risk assessment, and preventive measures
- *Holistic* – including beneficial preventative measures to multiple climate hazards, such as decentralised and renewable energy generation onsite, durable materials, appropriate landscaping and surface treatments to decrease heat,

flood and drought impacts, usage of green roofs and walls, tree planting, biomimicry water system design, usage of transitional spaces where inside meets outside by pergolas, planting areas, and so on to increase shading and temper of air before entering inside

In northern Europe, the city of Malmö in Sweden and its construction industry have created a local climate neutral roadmap to reach by 2030 (LMF30), and climate positive beyond 2030. It consists of six strategic and indivisible focus areas as follows:

- *Business models with incentives and collaboration* – Including green financing solutions, open emission reports, and supply-chain requirements regarding climate neutrality
- *Circular economy and resource efficiency* – Including valuation and selection of materials from circularity and climate impact factors point of view, full attempt to reach total circularity, as well as circular and bio-based construction methods
- *Design, process, and climate calculation* – Including LCA-based climate calculation during the construction process collaborating with stakeholders towards a sustainable architecture with integrated energy and ecosystems, mobility- and material-optimised solutions, and socially based solutions
- *Climate-neutral construction materials* – Including circular and climate-neutral material optimisation, materials with climate information by EDPs, prioritising circular and climate-neutral materials in the framework and foundations of a building, and validating climate information from the supply chain regarding climate information by EPDs
- *Operation management and maintenance* – Including implementation of minimised climate impact during operation and maintenance; using climate calculation to decide when and how to maintain, repair, exchange, or rebuild a building; aiming to reach 100% renewable energy for heating and operation; using green or sustainable leasing or renting contracts
- *Climate neutral construction sites and transports* – Including climate neutral construction with climate-optimised transport to, from, and within the site based on climate calculations, reducing the total need for transportation through efficient logistics, pooled deliveries, high rate of industrialised methods, and other mobility solutions; climate neutral construction sites by 2030

Critical path tools

Critical path tools aim to promote the process of construction projects and sustainability. It offers clients and design teams a way to manage outcomes by emphasising the positive environmental, social, and economic opportunities throughout the process. Generally, little attention is given to a construction project by a certification scheme for issues associated with the design, construction, and handover processes. However, conceptual design and sustainability objectives are particularly vulnerable to project costs. It is important that the design objectives are validated on completion before handover. Process tools are needed to support the management of the design and construction process and to reliably achieve objectives through successful project delivery and beyond into long-term sustainability. Project outcomes include improved control, minimisation of adverse environmental impacts, and verification of documented records delivered to the client during handover. Tender strategies, project management, cost-cutting issues, and building operations can potentially undermine a project's sustainability objectives. Sustainability is vital to the project agenda as the project progresses and an increasing number of documents provide real-time guidance. During the construction process, it is important to record several project objective issues regarding sustainability, such as:

*   Resource productivity and comparing multifunction and reuse options
*   Preparation of tender requirements with respect to sustainability objectives
*   Maintaining records and documentation of required sustainability objectives
*   Onsite training in sustainability and unusual elements of design
*   Assessing changes during the project process to minimise adverse impacts
*   Ensuring that the project outcome, including changes, is documented to deliver at handover
*   Training and involvement of operation organisation and users

The RIBA Sustainability Outcomes Guide by the Royal Institute of British Architects and The GlobalABC 'Roadmap for Buildings and Construction 2020–2050 – Towards a zero-emission, efficient, and resilient buildings and construction sector by the Global Alliance for Buildings and Construction' (see reference list) are examples of outlines of critical steps required to ensure that the process of sustainable design is correctly followed by all involved. The former is from the perspective of architects and designers, while the latter is aimed equally at the involved stakeholders. The examples above provide a roadmap rather than exact rules and requirements.

Building logbooks are an example of critical path guidance. Intended to combine the design team and the client's operation management, they convey how the building performs and provide a log of ongoing performance and maintenance. Considered essential tools to promote energy- and carbon-efficient operations through improved understanding, management, and operation, they also result in buildings with lower operating costs and maintained sustainability. Providing and contributing to enhanced occupant comfort, satisfaction, and productivity, logbook information is also beneficial to the users of the building. The logbook could be an easily accessible local source of current information for all those working and living in a building. An entry point for applying circular economic principles to the built environment, it uses sustainability indicators to measure the carbon, materials, water, health, comfort, and climate change impacts throughout a building's full life cycle.

Level(s) is the European Commission's framework to consolidate sustainable building thinking across the European Union by offering guidance on the key areas of sustainability in the built environment and how to measure them during design and after completion. It provides a common language for assessing and reporting the sustainability performance of buildings and promotes the use of lifecycle assessment (LCA) and lifecycle costing (LCC) to help understand the environmental impacts of LCA in parallel with the most cost-effective approaches to reducing them by LCC. Level(s) are delivered in the form of user manuals and reporting templates. Each manual explains the sustainability concept, how to implement it, and how to record and measure the results. These documents help project design teams to focus on sustainability aspects and provide guidance on how to make accurate performance assessments. They also address clients and investors by providing information on how future costs and risks can be proven by extending flexibility and lifespans and enhancing the long-term value of assets.

At first glance, level(s) could be mistaken as a building certification scheme, but unlike such schemes, it is actually more of a set of tools to help stakeholders with the many aspects of interpretations of sustainability. It is based on six macro objectives:

1 Greenhouse gas emissions during a building's life cycle, including operational energy use and lifecycle carbon emissions
2 Resource-efficient and circular material life cycles, including bill of quantities, construction waste and materials, design for adaptability use, and deconstruction design
3 Efficient use of water resources, including operational water use
4 Healthy and comfortable spaces, including indoor air quality, time outside the thermal comfort range, lighting, and acoustic comfort
5 Adaptation and resilience, including protection of occupant health and thermal comfort, and protection from an increased risk of extreme weather and flooding events
6 Optimised lifecycle cost and value, including LCC estimations, value creation, and risk exposure regarding indoor air quality

In turn, these objectives are supported by between two and four measurable indicators, as well as by carbon emissions and costs; tools exist for measuring other aspects such as the likely waste streams from a site, how adaptable a design may be in the future, and just how recyclable it is. Three levels in the scheme synchronise with the workflow of a construction project:

- *Level one* – Provides guidance to inform stakeholders in the early stages of design. There are no metrics; a background reading on the subject brings an initial awareness of what is relevant and important in the construction project to the client and other involved stakeholders.
- *Level two* – About focusing increased concern on the areas identified as the main priorities and quantifying actuals in the technical designs to help make decisions based on data. Recommendations on international standards for use and methodology may be applied.
- *Level three* – Examines actual monitoring and feedback, whether it compares actual onsite waste recycling to the level two estimates or ongoing post occupancy monitoring of comfort, energy, water, or other building performance. This verifies what works in practice compared with the models.

7.5 Sustainability and regenerative certifications

Building certification schemes to achieve sustainability ultimately seek to advance the building sector towards ever-higher levels of improvement and transparency. Recent years have seen continued growth in the number of green/sustainable building certification standards, and more buildings than ever are being certified. Globally, major certifications, such as LEED, BREEAM, Passivehouse, DGNB, and WELL, continue to be widely used. However, there are also many regional and national level standards that are being applied, such as the CASA certification in Colombia, the Estidama Pearl Certification in the United Arab Emirates, Green Star in Australia, the LBC 4.0,

Standard (USA; with a global scope), and Greenmark in Singapore. There are also regional rating systems, such as the Global Sustainability Assessment System (GSAS) in the Middle East and the Green Building Index used in Southeast Asia.

Green buildings and sustainable building certifications focus on low and close to zero carbon emissions, and energy-efficient buildings are an important ingredient in achieving a low-carbon building stock. However, it must be based on common definitions in conjunction with mandatory building codes and progressive policies. Many of these schemes have been released, certifying buildings that achieve net zero based on performance data, including standards from the national green building councils. Examples of net zero schemes include LEED Zero from the United States, EDGE Zero Carbon (UK), NollCO$_2$ (zero carbon emission from Sweden GBC), DGNB International from Germany, and ILFI Zero Carbon (USA) certifications. The Green Building Council of Australia has released an updated Green Star tool that requires buildings to be net-zero, fossil fuel free, and 100% powered by renewables in order to score the highest possible 6 star rating.

A comparison of different rating and certification schemes illustrates deviations in the priority setting of sustainability and environmental objectives. The major building certification schemes that have been on the market for ten years or more, such as BREEAM, LEED, and DGNB, mostly highlight environmental data. Specifically, they highlight the energy performance of buildings. Such a narrow focus on the separate aspects of sustainability has also been critically examined. While criteria regarding water, energy, material, or indoor climate can be found in almost all certification schemes, aspects such as managing climatic responsive design, advanced sustainability efforts, or neighbourhood impacts are significantly less frequent. Moreover, social criteria are rarely explicitly stated as indicators in the schemes, and their implementation is minimised due to the low grades that can be achieved. They can mainly be found in neighbourhood manuals or city rating schemes, and not in ordinary building certification standards.

Hence, there are several alternative rating systems and newly updated existing rating schemes that focus more on the social dimension of sustainability, such as health and well-being, or contribution to society, rather than only building performance environmental data. Examples of such regenerative certification systems are the LBC, One Planet Living, and WELL.

Regenerative options

Some comprehensive certification schemes that focus more on assessing buildings in use contain more extensive requirements in relation to barley environmental issues. The requirements imply not only that the construction project itself is sustainable, but also that the site and context should be restored and, in some cases, regenerated to their original condition and beyond. These certification schemes require integrated design objectives supported by verification, documentation, and simulation for energy performance and indoor air quality controls. One such standard is the LBC by the International Living Future Institute in the United States, which differs from more common rating systems by evaluating actual performance after 12 months of use rather than the intended performance expressed in the design phase.

The LBC, as an example, does not contain any methodology, workflow, or tool to reach compliance but is structured around a set of imperatives that are supposed to be applied to either new or existing buildings, interiors, landscape, and infrastructure. Considered to be one of the most rigorous and demanding assessment schemes, it requires buildings to be net-zero carbon and energy, not only guaranteeing energy supply via renewables onsite but also providing its entire water supply and managing all of its grey and blackwater onsite, in addition to guaranteed access to nature and the incorporation of biophilic design. An international scheme, its approach is considered sufficiently flexible to be applied in various climate zones and countries with their own characteristics and national preferences. However, in addition to the barriers to sustainability standards, e.g. cost, lack of knowledge and experience, and time to gather the necessary information required for accreditation, the scheme and regenerative standards are also criticised for a series of other reasons. These include:

- Lack of clarity in relation to what is considered a positive or restorative contribution and how it can be assessed.
- Being questionable in terms of their efficiency
- Having aims too ambitious to be adopted by designers without systemic thinking and ecological frameworks
- Raising issues related to the scale of the proposed solution, which is normally too small to provide clear ecological benefits, as well as feasibility in relation to how they can be integrated into the existing context of cities and neighbourhoods.

One significant barrier related to the implementation of the LBC example is national codes and requirements, especially those related to water provision and discharge. In general, this is prioritised to a centralised supply where maintenance and control are assumed to be more cost-effective, making it difficult for construction projects to use onsite water and sewerage systems. Another barrier is water quality control, as many authorities, local as well as regional or national, require water entering building premises to be of drinking quality. Consequently, it could prevent greywater reuse for other purposes, such as toilet flushing or washing machines. With regard to energy supply, regulations often prioritise connections to the public grid, and there are no incentives for the installation of small and individual onsite renewable energy systems with surplus feedback to the grid. Barriers may also relate to regulatory bodies' risk aversion in relation to innovative approaches to design and construction in favour of existing and traditional practices. While stringent and outdated building policies can be a significant barrier for sustainable and innovative construction, obstacles can also include a lack of clear codes and directives that would help organisations attract investment in green technology businesses and, hence, develop the industry further.

These regulatory barriers were mainly documented in the North American context through an analysis of case studies from the United States and Canada in relation to mandatory regulatory requirements for project approval, from interviews with project teams as well as other industry stakeholders. It also seems that there are no documenting regulatory challenges and barriers to implementing regenerative standards in the European context and potential overlaps between EU regulatory frameworks and the LBC scheme, which could make parts of the latter redundant and, hence, prevent its implementation in the European scenario.

The LBC scheme and regenerative principles

According to the 12 principles of regenerative design mentioned in Chapter 6, in Section 6.1, regenerative design is divided into topics such as place, energy, water, wellbeing, carbon, resources and waste management, equity, education, economics, culture and community, and environment and mobility. For each topic, some aspects of interest can be used to determine how rating tools and schemes can incorporate them. Below are the topics mentioned and their assessed regenerative goals regarding the level of quantifiable coverage.

- Place, culture and community, environment, and mobility:
  o regenerative land use
  o local community agriculture
  o biodiversity
  o community connectivity
  o natural-based solutions design
  o regenerative heritage
  o mobility

Regenerative aspects related to place, culture, and community focus on the integration of the sites into local natural and urban communities. The specific objectives include the incorporation of the design principles of bioclimatic, biomimicry, biophilic design, regenerative land use options, and community connectivity.

The LBC first imperative of the place petal focuses on environmental protection, that is, construction can only be undertaken on previously developed land, grey fields, or brown fields and has to be sufficiently apart from environmentally protected areas such as wetlands and old-grown forests, unless specifically connected to the protection of these sites. Furthermore, it protects prime farmlands but considers monoculture agriculture to be a suitable place for development. The local community agriculture aspect requires construction projects to incorporate areas dedicated to urban agriculture and/or medicinal plants and be kept, with two weeks of self-sufficiency for residential developments, unless the project's main purpose is to protect the land where it is being developed or restored to its original conditions. The biodiversity aspect focuses on the protection and expansion of existing natural areas and demands protection of biodiversity outside the project site through investments in approved conservation organisations. Regarding the environment and mobility, construction projects must provide specific measures to encourage building environments that support walking and cycling, and reduce the use of vehicles onsite.

- Energy
  o non-polluting energy sources
  o onsite renewable supply
  o net positive energy
  o onsite storage for resiliency
- Water
  o net zero water use
  o local stormwater management
  o wastewater treatment onsite without chemicals

Regenerative energy and water use imply a net positive water and energy use. The definition of these topics could be directly derived from the LBC scheme, as this certification is leading the way to regenerative energy and water use. The LBC energy petal imperatives are energy + carbon reduction and net positive carbon which include requirements for reduction and dependence on renewable sources and development of decentralised and safe infrastructure. More specifically, it requires new buildings to be 70% less energy intensive than the equivalent baseline building type, with 20% of it being associated with embodied energy in construction. In addition, onsite renewable energy production needs to cover 105% of annual energy use.

The LBC water petal imperatives responsible for water use and net positive water require water usage not to exceed the carrying capacity of the natural and local hydrologic cycle. In addition, the water footprint should be 50% lower than the regional baseline. This implies that the construction project needs to verify that it can supply the building with onsite sourced water for 12 months with monthly metering to verify the different supply sources. Moreover, the overall target water footprint must not exceed the natural capacity of the site at 50% of the regional baseline.

This petal also requires onsite stormwater, greywater, and blackwater management as well as onsite water harvesting from rain, underground, condensation, surface sources, and recycling for grey and blackwater after onsite treatment for reuse. Connecting to the potable-water grid is only allowed if the municipality enforces it or if the local aquifer is inaccessible or contaminated. However, the connection to the grid can only be performed for potable usage within the carrying capacity of the site. Connecting to the municipal sewer grid is allowed, but the treated water must return to the site for usage as long as the municipality has biologically based treatment.

- Carbon
  o net zero lifecycle CO_2 emissions
  o carbon positive impact technologies

Closely connected to the objectives of energy and water is a carbon topic that targets zero carbon emissions, not only in the use phase but also during the entire life cycle of a construction project and its outcome (see Section 7.1 in this chapter). Beyond the outcome of carbon emissions directly from the project life cycle, it is possible to use carbon storage techniques to increase the positive impact of the project on the climate.

- Well-being
  o working conditions connected to nature
  o indoor air quality
  o biophilic design
  o water quality
  o healthy food
  o accessibility
  o design for active life style
  o visual comfort
  o thermal comfort
  o acoustic comfort
  o mental health
  o medical support

The well-being topic of regenerative construction projects focuses on the comfort and health of the building occupants, similar to the WELL certification system, which is considered the most advanced rating tool for this topic. The LBC petal

Health + Happiness comprises three imperatives: a healthy interior environment, healthy interior performance, and access to nature. The first transfers central operation control to occupants, guaranteeing access to fresh air and daylight for a minimum of 75% of all spaces regularly used. The second set of indoor air quality targets mainly refers to standards for ventilation and a set of indoor air quality tests in compliance with current local test method standards before and nine months after occupation; however, no specifications are provided for indoor air temperature and humidity control. The third refers to promoting human–nature interactions through biophilic design approaches and making post-occupancy evaluations within 6–12 months of occupancy.

- Resources
  o material transparency
  o elimination of toxic materials
  o design for disassembly
  o responsible sourcing

The resource topic defines regenerative resource management with a lifecycle approach, which includes responsible sourcing, transparent reporting of built-in materials, elimination of toxic materials, and inclusion of options for disassembly during design.

The LBC Materials petal has five imperatives:

- Responsible materials require the construction project to have a minimum of 50% of the building's timber and wood material to be FSC-labelled, salvaged, or harvested on site during clearing the site for construction or to restore or maintain ecological function of the site. The remainder must originate from other low-risk sources with content, origin, and fair extraction processes proven as well as divert 80% of construction waste from landfills. This is also a requirement for using local sources of materials.
- The red list prohibits the use of dangerous materials, i.e. materials with the greatest impact on human and ecosystem health, such as asbestos, bisphenol A, chlorinated polymers including PVC, chlorofluorocarbons CFC and HCFC, PFCs, PCBs, toxic heavy metals such as lead, mercury, arsenic, cadmium, and volatile organic compounds.
- Responsible sourcing requires purchasing materials from approved suppliers or supply chains, third party labelled
- Living economy sourcing requires the use of materials a short distance from the project site
- Net positive waste focuses on durability, adaptability, and reuse of new materials to be able to deconstruct with adaptable reuse of materials and with re-integration to the nutrient loop at the end of material life. Materials must be diverted from landfills.

- Equity
  o diverse, inclusive users
  o accessibility
  o investment in local/global community

o integration of cultural heritage
o transparency of company procedures
o regenerative CSR programmes

The equity topic targets building users through design and operation goals for inclusivity, accessibility, transparency, and investment in local/global communities as well as CSR programmes and cultural heritage.

The LBC equity petal is centred on promoting human-scale rather than vehicle-scale places in universal access and inclusion, focusing mainly on restricting the size and percentages of land use for car parking. It also focuses on creating places with universal accessibility (regardless of gender, age, socioeconomic class, etc.), including universal access to nature, as well as outdoor fresh air and solar rights, compared with biophilic design approaches. Lastly, it focuses on equitable investment, requiring 0.1% of the total project cost to be donated to charity, and requests transparency and disclosure of business practices of the construction project team towards promoting an equitable society.

• Education
   o participatory processes
   o inspiration/education

The education topic also targets building users and a wider community, with the objectives of wide participation and inclusion of education programmes.

The LBC education topic is included in the beauty petal, as the imperative of Education + Inspiration is aimed at providing educational materials about the operation and performance of the construction project to the occupants and the public to share successful solutions and catalyse a broader change. All projects must provide a project case study, an annual open day for the public, and a copy of the operations and maintenance manual. Moreover, all projects except residentials must provide a short brochure describing the design and environmental features of the project, install interpretive signage to teach visitors and occupants about the project, and develop and share an educational website about the project.

The other imperative of beauty, a petal concerning qualitative measures which includes design that uplifts the human spirit, is Beauty + Biophilia, which implies connecting project teams and occupants with the benefits of biophilia and incorporating meaningful biophilic design elements into the project. Construction projects must be designed to include elements that address the inborn human/nature connection. Each project team must engage in a minimum of one whole-day exploration of the project's biophilic design potential. Exploration must result in a biophilic framework and plan for the project that outlines strategy and implementation of how the project will be transformed by deliberately incorporating nature through environmental features, light and space, and natural shapes and forms (a biomimicry approach). Moreover, the project will be transformed by deliberately incorporating nature's patterns through natural patterns and processes and evolved human–nature relationships. The project will be uniquely connected to the place, climate, and culture through place-based relationships; a meaningful integration of public art; and containment of design features intended solely for human delight and the celebration of culture, spirit, and place appropriate to the construction project's function.

- Economics
  - o participation in sharing economy
  - o restorative enterprise
  - o building circular economic value chain

Economics focuses on the integration of a project into the circular economy value chain. The link between the sharing economy and the built environment is also considered. There is no particular LBC petal about economics; instead, it is blended with the other petals, especially with the resource petal.

Level of regenerative indicators

Regarding the benchmarks for the defined regenerative indicators, quantifiable targets for certain regenerative aspects are identified. In the case of energy, water, and carbon use, net positivity can serve as a regenerative benchmark. In other cases, there is no clear limit between the sustainability and regenerative objectives, such as to determine the regenerative comfort parameters. In addition, objectives where the level of compliance cannot be quantified (e.g. in the case of the objectives of biophilic design and beauty) have to use qualitative valuation and verification. Because of the various types of benchmarks, the relation of the strictness of requirements in the different assessment schemes to the regenerative indicators differs by quantifiable or qualitative evaluation, such as if the contribution of the requirement is unquantifiable (e.g. the LEED Integrative Process credit requires attempting an energy optimisation of the construction project, which cannot be translated to a quantifiable value). Another example is when the credit has lower benchmarks than the regenerative benchmark (e.g. in the LEED assessment scheme, the maximum points can be achieved by improving the proposed building performance rating by 50%, compared with the baseline). Finally, if the credit requires the same strictness as regenerative benchmarks (e.g. the LBC assessment requires net positive energy use) or it represents the most stringent values possible (e.g. according to WELL, PM2.5, concentrations less than 15 μg/m indoors).

However, LBC and WELL target stricter levels of comfort, material sourcing, and transparency than the more traditional sustainability rating tools, implying that they not only limit the negative effects of an artificial environment but also attempt to implement positive effects, such as improved health and productivity.

Regenerative economics is minimally covered in the different systems, but DGNB has an alignment of circular economy principles. Other systems refer mainly to regenerative economic targets, including indicators requiring participation in the sharing economy (e.g. sharing community spaces with local communities or shared transport facilities). Regarding the depth of alignment with regenerative objectives, LBC targets positive impacts on all the regenerative objectives in four topics (place, energy, water, and carbon). In two other categories – resources and education – LBC covers all topics, but the requirement could be more comprehensive (e.g. participatory project development is only partially included). In the well-being and equity categories, some of the targets are not addressed in the system, and the economic category is not explicitly covered. Regarding the other rating tools, WELL incorporates all regenerative well-being objectives and partially covers only four other topics. DGNB and BREEAM incorporate all categories, but in most cases, not all aspects, and with less comprehensiveness. LEED does not cover equity or education topics.

An analysis by Andreucci et al. (2021) of the five selected rating tools discussed above (BREAM, LEED, DGNB, WELL, and LBC) regarding sustainability or regenerative content in the certification schemes suggests that the schemes, based on their purpose (sustainability, wellness, and regenerative systems), are the defining factor of how they incorporate regenerative objectives. As expected, because it was developed specifically to address regenerative development, LBC is much more comprehensive in its incorporation of the identified aspects, addressing most issues highlighted in the academic literature. On the other end of the scale, WELL is more narrowly focused, as it provides healthy buildings for occupants, but by incorporating the widest range of health-related regenerative aspects possible. In the middle range, the main sustainability-oriented rating tools, such as LEED, BREEAM, and DGNB, provide good coverage of the regenerative objectives that originated from traditional sustainability objectives, but with limited coverage of education and equity topics. Among the three latter systems, LEED does not cover equity and education topics at all, nor does it set comprehensive targets in the other categories.

Bibliography

Andreucci, A.M. et al. (eds.) (2021) *Rethinking Sustainability Towards a Regenerative Economy, Future City*, Vol 15, Springer Nature, Cham, Switzerland, 2021.

Baper, S.Y. et al. (2020) *Towards Regenerative Architecture: Material Effectiveness*, International Journal of Technology 2020, 11(4), 722–731.

BASTA online (2022) BETA to BASTA, BASTA, https://www.bastaonline.se/how-it-works/beta-to-basta/?lang=en, access 2022-04-11.

Boverket (The Swedish National Board of Housing, Building, and Planning) (2019) Grafiskt material för livscykelanalys av byggnader, https://www.boverket.se/sv/byggande/hallbart-byggande-och-forvaltning/livscykelanalys/grafiskt-material/, access 2022-03-18.

Boverket (The Swedish National Board of Housing, Building, and Planning) (2020) *Regulation on Climate Declarations for Buildings – Proposal for a Roadmap and Limit Values, Report 2020:28*, Boverket, Karlskrona, Sweden, 2020.

Boverket (The Swedish National Board of Housing, Building, and Planning) (2022) https://www.boverket.se/en/start/building-in-sweden/developer/rfq-documentation/climate-declaration/questions/, access 2022-03-16.

BREE (2022) The Green Guide to Specification, https://www.bregroup.com/a-z/the-green-guide-to-specification/, access 2022-03-28.

BREEAM (2022) Building Research Establishment Environmental Assessment Method, https://www.breeam.com, accessed 2022-04-01.

Cabeza, L.F. et al. (2022a) Buildings. In IPCC, 2022: Climate Change 2022: Mitigation of Climate Change. Contribution of Working Group III to the Sixth Assessment Report of the Intergovernmental Panel on Climate Change [P.R. Shukla, J. Skea, R. Slade, A. Al Khourdajie, R. van Diemen, D. McCollum, M. Pathak, S. Some, P. Vyas, R. Fradera, M. Belkacemi, A. Hasija, G. Lisboa, S. Luz, J. Malley, (eds.)]. Cambridge University Press, Cambridge, UK and New York, NY.

Cabeza, L.F. et al. (2022b) Buildings Supplementary Material. In IPCC, 2022: Climate Change 2022: Mitigation of Climate Change. Contribution of Working Group III to the Sixth Assessment Report of the Intergovernmental Panel on Climate Change [P.R. Shukla, J. Skea, R. Slade, A. Al Khourdajie, R. van Diemen, D. McCollum, M. Pathak, S. Some, P. Vyas, R. Fradera, M. Belkacemi, A. Hasija, G. Lisboa, S. Luz, J. Malley, (eds.)].

CATS (2022) The Transect, Centre for applied transect studies, https://transect.org/transect.html, access 2022-04-06.

City of Melbourne (2022) Sustainability checklist and factsheets, Increase the climate resilience of your building – Fact sheet, https://www.melbourne.vic.gov.au/building-and-development/sustainable-building/Pages/sustainability-checklist-fact-sheets.aspx, access 2022-03-28.

Clark, G. and Tait, A. (eds.) (2019) *RIBA Sustainable Outcomes Guide*, Royal Institute of British Architects, RIBA, London, UK, 2019.

Cradle to Cradle (2022) What is cradle to cradle certified? https://www.c2ccertified.org/get-certified/product-certification, access 2022-04-10.

DGNB (2022) Deutsche Gesellschaft für Nachhaltiges Bauen, https://www.dgnb-system.de/en/system/, access 2022-04-01.

Dodd, N. et al. (2021) Level(s) – A common EU framework of core sustainability indicators for office and residential buildings, User Manual 1: Introduction to the Level(s) common framework (Publication version 1.1,) JCR Technical Report, Joint Research Centre, Directorate B, Growth and Innovation Unit 5, Circular Economy and Industrial Leadership, European Commission, Seville, Spain, 2021.

EN15978:2011 (2011) *Sustainability of Construction Works – Assessment of Environmental Performance of Buildings – Calculation Method*, European Committee for Standardization, CEN, Brussels, Belgium, 2011.

EU Ecolabel (2022) EU Ecolabel, European Commission, Directorate-General for Environment, https://ec.europa.eu/environment/ecolabel/, access 2022-04-10.

EU Energy Label (2022) Energy label and eco-design, European Commission, Directorate-General for Communication, https://ec.europa.eu/info/energy-climate-change-environment/standards-tools-and-labels/products-labelling-rules-and-requirements/energy-label-and-ecodesign_en, access 2022-04-10.

Fairtrade Marks (2022) The fairtrade marks, Fairtrade International, https://info.fairtrade.net/what/the-fairtrade-marks, access 2022-04-10.

Forsberg, M. and Bleil de Sousa, C. (2021) *Implementing Regenerative Standards in Politically Green Nordic Social Welfare States: Can Sweden Adopt the Living Building Challenge?* Sustainability 2021, 13, 738.

FSC (2022) Forest management certification, Forest Stewardship Council, FSC, https://fsc.org/en/forest-management-certification, access 2022-04-10.

GlobalABC/IEA/UNEP (Global Alliance for Buildings and Construction, International Energy Agency, and the United Nations Environment Programme) (2020) *GlobalABC Roadmap for Buildings and Construction: Towards a Zero-Emission, Efficient and Resilient Buildings and Construction Sector*, IEA, Paris, 2020.

Haliday, S. (2019) *Sustainable Construction*, 2nd edition, Routledge, New York, NY, 2019.

Haselsteiner, E. et al. (2021) *Drivers and Barriers Leading to a Successful Paradigm Shift toward Regenerative Neighborhoods*, Sustainability 2021, 13, 5179.

ILFI (2019) Living Building Challenge 4.0 – A visionary path to a regenerative future, International Living Future Institute, Seattle, WA, 2019, https://living-future.org, access 2022-04-04.

IPCC (2022) Summary for Policymakers. In Climate Change 2022: Mitigation of Climate Change Contribution of Working Group III to the Sixth Assessment Report of the Intergovernmental Panel on Climate Change, [P.R. Shukla, J. Skea, R. Slade, A. Al Khourdajie, R. van Diemen, D. McCollum, M. Pathak, S. Some, P. Vyas, R. Fradera, M. Belkacemi, A. Hasija, G. Lisboa, S. Luz, J. Malley, (eds.)]. Cambridge University Press, Cambridge, UK and New York, NY, https://www.ipcc.ch/report/ar6/wg3/, access 2022-04-10.

IWIBI (2022) International WELL Building Institute, https://www.wellcertified.com/certification/v2/, access 2022-04-04

LEED (2022) Leadership in Energy and Environmental Design, https://www.usgbc.org/leed, access 2022-04-01.

Lehmann, H. (ed.) (2018) *Factor X – Challenges, Implementation Strategies and Examples for a Sustainable Use of Natural Resources, Eco-Efficiency in Industry and Science*, Vol 32, Springer Publishing, Cham, Switzerland, 2018.

Level(s) (2022) European framework for sustainable buildings, https://ec.europa.eu/environment/levels_en, access 2022-04-08.

LFM30 (2019) How we collectively develop a Climate Neutral Building and Construction Industry, https://lfm30.se/wp-content/uploads/2021/01/Local-Roadmap-LFM30-English.pdf, access 2022-03-28.

Lwasa, S. et al. (2022) Urban Systems and Other Settlements. In IPCC, 2022: Climate Change 2022: Mitigation of Climate Change. Contribution of Working Group III to the Sixth Assessment Report of the Intergovernmental Panel on Climate Change [P.R. Shukla, J. Skea, R. Slade, A. Al Khourdajie, R. van Diemen, D. McCollum, M. Pathak, S. Some, P. Vyas, R. Fradera, M. Belkacemi, A. Hasija, G. Lisboa, S. Luz, J. Malley, (eds.)]. Cambridge University Press, Cambridge, UK and New York, NY.

Natureplus, e.V. (2022) Certified sustainable, International Association for Sustainable Building and Living – natureplus e.V, https://www.natureplus.org/index.php?id=17&L=2, access 2022-10-11.

Nordic Swan Ecolabel (2022) Sets of criteria, Nordic Ecolabelling, https://www.nordic-ecolabel.org/product-groups/, access 2022-04-11.

Overshoot day (2022) About Earth overshoot day, https://www.overshootday.org/about-earth-overshoot-day/, access 2022-04-10.

Reed, B. (2007) *Forum: Shifting from 'Sustainability' to Regeneration*, Building Research & Information 2007, 35(6), 674–680.

Spangenberg, J.H. et al. (1998a) *Material Flow Analysis, TMR and the MIPS Concept: A Contribution to the Development of Indicators for Measuring Changes in Consumption and Production Patterns*, International Journal of Sustainable Development 1998, 2(4), 491–505.

Spangenberg, J.H. et al. (1998b) *Material Flow-based Indicators in Environmental Reporting Environmental Issues Series Report 14- Material Flow Analysis Method*, European Environment Agency, Copenhagen, Denmark, 1999.

UKWAS (2022) UKWAS Standard, https://ukwas.org.uk, access 2022-04-10.

UNEP (2020) *2020 Global Status Report for Buildings and Construction: Towards a Zero-emission, Efficient and Resilient Buildings and Construction Sector*, United Nations Environment Programme Nairobi, Kenya, 2020.

Wackernagel, M. and Rees, W. (1998) *Our Ecological Footprint – Reducing Human Impact on the Earth, The New Catalyst Bioregional Series*, 9, New society Publishers, 6th printing, Gabriola Island, B.C Canada, 1998.

WGBC (World Green Building Council) (2022) https://www.worldgbc.org, access 2022-03-16.

8 Further on – construction for a regenerative future

8.1 The basics: regenerative development

Most construction projects address the efficiency of the product or the building through a mainstream, green, or sustainability concept, while failing to understand the systems they are trying to sustain. From the formulation of the UN SDGs and roadmaps towards the environmentally crucial period between 2030 and 2050, the subject of regenerative development is gaining more attention in the construction sector. Regarding the urgent necessity to address ensuing climate change threats, it is time to change this mental attitude to one that better reflects how our environment truly works and enables us to design and build with the whole system in mind. Whole systems and living systems thinking, which can help transform the way of practising sustainability towards regenerative development to linking the natural and built environments in a holistic manner, is crucial.

Reed (2007) describes the three levels that pave the way towards regeneration as 'to widen the view from only seeing the trees in the forest (level 1) to seeing the whole forest from outside (level 2) and, lastly, from above, discovering that there are a lot of other forests around (level 3)'. When transforming this into a construction concept, Level 1 concerns efficiency, such as energy efficiency or material productivity. Level 2 relates to sustainability as the effectiveness of the UN SDGs or the triple bottom-lines, and Level 3 concerns understanding the entire system, as shown in Figure 8.1 (similar to Figure 1.19). The learning process evolved from the pre-level conventional practice of not breaking regulations or building codes to high performance as a technically efficient approach. Green is considered a general term implying a direction of continual improvement toward an ideal of 'doing no harm' with zero impact on the environment. This idea is called sustainability.

Beyond sustainability, an entire system mindset begins with a 'doing more good' approach. The first step, a restorative stage of approach, is to use the activities of design and building to restore the capacity of local natural systems. The next step, reconciliation, involves humans as an integral part of local nature, considering human and natural systems as one. Regenerative development is a process that engages and focuses on the entire system, of which humans are a part. Humans can participate in places, communities, watersheds, and bioregions. By engaging all the key stakeholders and processes of the place, humans, other parts of biotic diversity, earth systems, and the consciousness that connects them, the process builds the capability of humans and the 'more than human' participants to engage in a continuous relationship. The process supports continuous learning through feedback, reflection, and dialogue so that

DOI: 10.1201/9781003177708-8

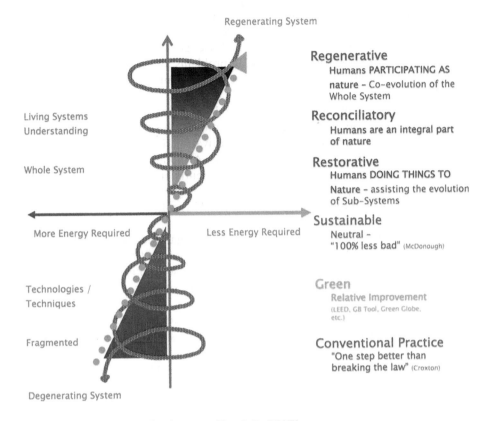

Regenerating System

Regenerative
Humans PARTICIPATING AS
nature – Co–evolution of the
Whole System

Living Systems
Understanding

Reconciliatory
Humans are an integral part
of nature

Whole System

Restorative
Humans DOING THINGS TO
Nature – assisting the evolution
of Sub–Systems

More Energy Required   Less Energy Required

Sustainable
Neutral –
"100% less bad" (McDonough)

Technologies /
Techniques

Green
Relative Improvement
(LEED, GB Tool, Green Globe,
etc.)

Fragmented

Conventional Practice
"One step better than
breaking the law" (Croxton)

Degenerating System

Figure 8.1 Regenerative development (Reed, B. (2007)).

all aspects of the system are an integral part of the process of life in that particular place to maintain sustainability.

Regenerative development is the place-based development of capabilities necessary for living systems to increase complexity, diversity, and capacity to support all life and provide future options, such as health and well-being.

Regenerative sustainability

Regenerative sustainability represents a necessary worldview and paradigm shift that includes and exceeds conventional sustainability by adopting a holistic worldview. Regenerative sustainability, as described above regarding regenerative development, views humans and the rest of life as one system with continual processes to consolidate the uniqueness of each place or community. The aim of regenerative sustainability is to consolidate living systems in a fully integrated individual-to-global system. This requires humans to adapt to living system principles of wholeness, change, and rela-tionships, as nature does. The inhabitants of a place or community, along with the stakeholders who developed it, ultimately decide whether it is (un)sustainable or (not) thriving, but places are constantly changing. Developing communities that can continually support higher levels of health and well-being are necessary for regeneration. These abilities include adaptation, self-organisation, and continual improvements, as well as decisions about infrastructure, land use, governance, food systems, cultural practices, and lifestyles. Instead of seeing problems and solutions, regenerative sustainability views living systems as existing under changing conditions of health and complexity.

Regenerative sustainability may be more difficult to implement than conventional sustainability because of its more ambitious aims; however, studies imply that regenerative approaches are more inspiring and motivational than conventional approaches and are more effective at achieving these aims. Although the concept of regenerative sustainability has been articulated relatively recently, its principles have been applied in practices such as regenerative development, ecological and regenerative design, ecological planning, and regenerative agriculture. This offers insights into what is possible when adopting a holistic view.

Ways of adopting regenerative sustainability are gaining more attention in relation to the shifting role of buildings and the changing construction process due to climate change. Such change is still in its infancy and poses several practical and operational concerns for clients and construction project teams, such as:

- Living systems thinking represents too great a leap for mainstream construction.
- The additional time required in the early design stages to engage stakeholder input and client commitment is a major obstacle.
- The scale of construction projects at which most project teams work might be incompatible with regenerative ambitions.

Only a few regenerative development case studies have been published so far, and most are in rural areas with small and affluent community groups. Urban and suburban areas are not homogenous in terms of diversity in social and economic equity, demographics, and closeness to services. Different neighbourhood communities within cities can also have qualitatively different engagements with place, relations, and capabilities. Regenerative development can be considered inappropriate in complex urban situations. Unless more case studies are conducted in these contexts, regenerative applications may be restricted to the outer edges of urban areas.

Regenerative community

Communities are defined as biotic (living factors that influence the environment) and abiotic (non-living factors that influence the environment) components of the interaction of the complex webs of life. This contrasts with the neighbourhood, which comprises a geographically bounded area in a human-dominated system. Communities are blocks of nature and societies, neighbourhoods, cities, landscapes, watersheds, and bioregions. Working with the structure of living systems at a community level could stimulate regeneration across all scales; for example, community action at the neighbourhood level stimulates change at the city level, which stimulates changes at the landscape and bioregional level. Research and action at the community level are key to sustainability and regeneration and should be holistic. Its processes and tools help inhabitants better understand life-giving flows, such as water, food, energy, organisms, and information, through their community and the communities of which they are a part, as well as their relationships. It transforms living system principles and characteristics into general indicators and strategies for whole-system health and regeneration, specific to a place through ongoing processes. These processes help the inhabitants and stakeholders of a place to integrate ecological and sociocultural dimensions of living systems, as well as development, such as participatory processes in urban design and landscape architecture, and products (e.g. ecological urban infrastructure, city plans and codes, buildings, and water systems) that are involved in community development. In short, regenerative community development seeks to develop regenerative communities that form the matrix, out of which all aspects of regenerative beings, regenerative living, and regenerative whole living systems arise.

Thousands of ecovillage communities around the world that have been developing ecologically and socially in recent decades are of particular interest. Although ecovillages most often fall into the category of regenerative design, many are attempting to shift to larger communities, of which they are a part, to create a bigger social change. They are living laboratories that provide years of data and invaluable insights into sustainable and regenerative living. The UN recognises the ecovillage as the best strategy for achieving sustainability goals. Ecovillages can provide a springboard for increasing community development efforts to become regenerative on a larger scale.

Regenerative buildings

The differences between green and regenerative approaches are significant. Current green approaches already have many practical experiences, whereas regenerative approaches are undertaken without sufficient track records to fully support the goals. As previously mentioned, there is an urgent need to reduce global carbon emissions during the decades leading up to 2050. Thus, the design stage of the construction process, focusing on restricting and totally reducing carbon emissions, will dominate future work, through proactive efforts, shifting client demands and expectations, or more stringent building codes. However, as with past environmental issues, a focus on climate change with the seriousness it deserves will likely be compromised by other, more locally pressing societal and political priorities. Furthermore, the construction process itself is unable to fundamentally shift to a regenerative approach. New strategies and contexts can probably be implemented partially and selectively within an already existing performance, depending on the involved parts' experience, capability, and commitment in zero-or beyond zero carbon buildings. However, the necessary reductions in carbon emissions from the built environment have not yet been generated by green building strategies, and environmental performance has improved in recent decades.

As new infrastructures and buildings become more resilient to a changing climate, adaptation to regenerative development will depend on people's day-to-day actions in the places in which they live. Unlike green building practices, regenerative approaches assign design professionals as co-learners and co-creators, together with community members and other stakeholders, that is, those who will be most directly affected by climate change. However, many green building design strategies and technical knowledge remain valid even when other environmental impacts are reduced. As regenerative development gains more acceptance, green building knowledge and experience will need to be revised into more holistic thinking and a better understanding of interactions between strategies and many qualitative factors.

Regenerative development and design are currently in the same position as that of green buildings when they emerged some decades ago; however, they are unlikely to become mainstream in the same way or as fast. The support of non-governmental organisations, such as the Green Building Councils, must be revised to achieve the same level as that of green buildings. Of course, this will rely on the direct engagement and involvement of community members and stakeholders, who are probably more framed within an ecological worldview and adaptable to whole system thinking, while clients are not. Clients need to reconsider their task of managing and operating their building stock from a regenerative view and approach.

8.2 The client and the regenerative process

Shifting from a sustainability focus to a generative process requires a level of commitment from the client to diverge from conventional, linear design process management and to consider it as an opportunity to learn. A transition from Learning Level I, doing the same things in a better manner (efficiency), to Learning Levels II and III, which generate new levels of systemic understanding, is necessary. This learning process requires the project management team to engage deeply, participate, and be conscious of the Earth and human systems that are essential to the long-term health of the place and community. Thus, client and project management teams have become learning organisations. The client must shift to a regenerative process in the early inception phase of the construction process through their own initiative or that of the client advisor. As regeneration is a Learning Level III process, the frame of the project objectives must transition from imposing upon nature to a partnership with nature and acknowledgement of the deep relationship between humans and Earth systems. The aim is not merely to create a landscape and local habitat for a productive and healthy ecosystem; engagement with the entire system of what makes a place healthy is also required. From the perspective of understanding the entire system, an entry point can be found in any of its small or large parts. Each is an integral part of a living system and a key role can be identified for any system.

Regenerative processes are established through three essential aspects. Not ordinary steps, they involve continuous improvement over time; the process continually develops as the subject changes. The process must be intentionally sustained long after a construction project is complete. If not, the relationships established during the design phase will become diminished.

The three aspects include:

- Understanding the master pattern and story of the place
- Translating the patterns into design guidelines
- Developing a conceptual design with ongoing feedback: a conscious process of learning and participation through action, reflection, and dialogue

The first aspect requires an understanding of the type of human intention that the project aims to realise and, largely, the unique character of the place it seeks to inhabit. In contrast to conventional processes, gathering knowledge from different fields, such as water, energy, and soil, can be both fragmenting and misleading. The intentions of the client and other stakeholders must be identified, such as the drivers of the project and what is important to the client and the main stakeholders regarding the project and site. This should be determined through dialogue between stakeholders and expressed in qualitative and process terms.

To learn about the place, considerations should address the status of the ecosystem on site and the role of the project to take place inside of it, how the intentions of the project can be supported by the system, how it works, and why. One way is to study historic and present patterns in relation to human and ecosystem relationships, that is, between humans, plants, animals, hydrology, metrology, and geology. This can be described through general and approximate details, such as when and why life fully evolved and when, how, and why it changed. Carefully reading the landscape surrounding the project site and developing knowledge about key intersections where small interventions stimulate the entire system is necessary. The aim is to determine when the investment ensures a greater biodiversity yield than the actual impact of the project. Thus, to develop the story of the place, other important stakeholders should be engaged in the learning and understanding process of the project's relationship with complex surrounding ecosystems. The story of a place serves many purposes as context. First, the ability to sustain, implement, and maintain necessary changes, day after day, with care that comes from a deep connection to a place, requires a clear cultural relationship with the specific place. Second, learning about the story of a place enables an understanding of how its ecosystems work and, thus, how to achieve mutually beneficial results. Third, the story of a place provides an integrative context that helps to maintain a collective and meaningful purpose. Finally, the story of the place provides a framework for a continual learning process that enables humans to interact with their environment.

Once the desired story of a place and its pattern of relationships are defined, the task is to translate it into conceptual design and design guidelines, which serve as the framework for decisions made in the subsequent phases of the construction process: design and selection of appropriate green materials and technologies, construction, operation, and long-term operation and maintenance. In collaboration with the design team, the client establishes a development concept from an understanding of the story of the place, integrating project intentions, human needs, and the mutually beneficial relationship with the ecosystems of the site and its surrounding context. The story of a place is married with intentions for the future. This is the point in the regenerative process when the conceptual design begins.

The core team

To develop the regenerative process further, significant dialogue is required regarding the real issues of the environment and the intentions of the project within the limits of the place. At this point, it is important to form a core team to maintain these intentions in relation to the place and project. The core team is not responsible for day-to-day activities, but for remembering, maintaining, and promoting the core of the project and stimulating the resilience of the place and the project. The work of the core team is essential for implementing the regenerative processes. Without a core team monitoring project intentions and understanding the place, the process reverts to old patterns. When the initial design work is finished and the design team disbands, the remaining key stakeholders will need to sustain and develop a conceptual, conscious-learning, and feedback process in the future. An ongoing core team will ensure long-term continuous monitoring and measurement, with the involvement of stakeholders. The team will facilitate iterative cycles of action, reflection, dialogue, and learning as deepening connections, growing understanding, and mutual caring of the project within the place as it shifts from sustainability to regeneration.

The project organisation

Sustainability performance and its impact may be particularly important for clients. Additionally, corporate social responsibility may be important in the delivery of products in the built environment. Sustainability and regenerative requirements can be part of quality, resource, and procurement management. The requirements can be related to the product itself, the building, and the performance of the product, e.g. limitations of carbon emissions and energy consumption. In addition, requirements may be prescribed for any impact on the local topography or adjacent areas. Finally, outcomes in terms of the local community, such as providing employment and training opportunities or the use of local supply chains, may be determined.

From a sustainability and regenerative point of view, the project will be provided with a management framework for the planning and implementation of construction activities in accordance with the client's commitments, client organisation's EMS system, project context, end users, or any other stakeholders. When a regenerative approach is taken, it is important to implement the client's core team in the project's organisation, which will influence key design parameters related to sustainability, performance, and operational technologies. The overall management criteria for the project are also outlined, including the key success factors for project quality in terms of sustainability management.

8.3 Regenerative design

The idea of considering the entire surrounding built environment when designing a construction project – a regenerative design approach – instead of looking at the actual single building and its inherent components as a holistic approach, is a concept that promotes looking beyond sustainability and zero carbon emissions. A regenerative built environment is based on the uniqueness of a place, community engagement, and the continuous creation of a story of a place within the community. Concentrating only on building energy consumption or carbon emissions can risk shifting the environmental impact of a building from one factor to another. In contrast, regenerative design principles consider the built environment as a whole without sub-optimisations.

Regenerative design aims to achieve net positive development using a whole-systems approach that can generate synergies between ecological, economic, and social dimensions. The design approach considers the comprehension between the construction project and its site, place, and throughout the building's entire life. It requires a whole living systems approach, interconnectedness with nature, focus on the location and participation of all stakeholders, and enables continuous replacement, renewal, and rebirth. The entire construction process – building design, construction, and operation – positively influences the social, ecological, and economic health of the place within which the building exists, using the health of ecological systems as a basis for design that supports the regeneration of existing and lost ecosystems. The regenerative design approach creates a development that can restore health to both human communities and the ecosystems of which they are a part, and engages all stakeholders and processes of the place based on the co-evolution principle, that is, continuously learning how humans can participate in their surrounding environment. This design approach is more complex than ordinary construction design. Although it encompasses all green and sustainability criteria, the difference is that it replaces linear processes with cyclical ones, requires a shift from prescriptive to descriptive metrics, product-based to process-based, and linear to non-linear thinking, and involves a shift away from rules and regulations. This requires a qualitatively different framing of design that complements and supplies the currently available mainstream green building certification tools.

Once the desired patterns of the relationship between the construction project and its location, keystone species, and key system are understood in general terms, the metrics and ways to measure improvements can be established. One way to achieve this is to use the method of ecosystem services analysis (ESA) previously mentioned to identify indicators. However, this identification and assumed understanding are not a completely ensured truth of the whole, nor assurance that the stakeholders involved in the project will interact in the assumed way. Therefore, it is essential to continuously monitor the work to obtain the necessary feedback to allow the evolution of the system. This feedback process supports conscious engagement and deep relationships between people and places. The process of adapting all parts in relation to the entire process requires a few iterations of thinking. However, the method of designing a construction project is a linear process that depends on a temporal schedule, and it is necessary to make approximations simultaneously with the entire project through rapid and frequent iteration of ideas, as a process of integrative design.

The regenerative design process is iterative and requires improved involvement of project stakeholders during the design process, providing site-specific information and stories. Furthermore, stakeholders assess the construction project throughout its operation period and keep it regenerative.

The regenerative design process extends throughout the entire life cycle of the construction project, including the operation and construction phases. The design work and materials used need to be IT-intensive, for example, using BIM technology. The design must be from a 'glocal' point of view, reaching global targets with local adaptation and considerations. The design should be based on story- and place-specific parameters, intensive collaborations with the community, and feedback. The design team must use nature-based solutions as a guide to solve design problems and be interdisciplinary, consisting of members with different professional backgrounds (e.g. architects, civil engineers, mechanical engineers, computer engineers, electric engineers, biologists, and social anthropologists). The design team is responsible not only for an environmentally sufficient design but also for the regeneration of nature in the surroundings of the site. The team should have good coordination skills, as they must involve both the stakeholders and the community in the design process and in retaining knowledge of the IT-based technologies and materials used in the project to support the regeneration performance of the construction project. For the regenerative design process, it is necessary to establish an interdisciplinary design team to adapt contemporary technologies to the process.

The three design strategies

One regenerative design approach, described by Attia (2018), is based on three design strategies:

* Selection of a construction system
* Defining design elements and their performance
* Choice of regenerative materials

These three basic strategies should be applied at the start of the early conceptual phase of the design process and used throughout the entire process. The application of these strategies introduces a new design-thinking paradigm, in which sustainability is extensively integrated throughout the design process.

When applying the first strategy, the selection and sizing of the construction system are based on the concept of modularity, and the possibility of assembling and disassembling different regenerative materials and products. From a sustainability perspective, special attention should be paid to ensuring that the system be designed from a disassembling point of view to facilitate adaptability and structural flexibility, adding to the extended presumed lifetime of the facility. Based on the concept of a construction system, the entire envelope of a building can be developed with regard to flexibility and circularity. Furthermore, the envelope must satisfy the insulation and hygrothermal requirements of onsite performance conditions using renewable and net-positive impact materials.

Once the construction system has been chosen, the secondary strategy is to analyse the spaces of the building to evaluate nature-based solutions and connections to ecosystem services. Depending on the microclimate condition of the building site and its geographical location, certain design elements may be more appropriate as regenerative quality markers (e.g. atriums, courtyards, terraces, balconies, skylights, glazed facades, staircases, meeting rooms, open office spaces, common areas, foyers, and roof gardens). The integration of these elements provides quality and has a positive impact on end users. The purpose of using regenerative design elements is to improve indoor and outdoor air quality and water usage, health, and well-being of end-users, increase biodiversity, enable cultural and social cohesion, and generate energy. Defining these design elements depends on the type of construction system and scope of the building performance targets and indicators established by the client. Various key areas of interest, such as air quality, human health, energy savings combined with local renewable energy production, water management, and natural design, must be considered when designing a construction project.

When targeting a positive regenerative impact, it is important to improve the indoor and outdoor quality with regard to the operation of the building and the well-being of users. The latter is enhanced by the carefully designed availability of natural light and ventilated spaces for living and working in elements such as gardens, meeting rooms, and common spaces, including staircases. Especially in indoor environments, it is important to eliminate fine particles in the air, reduce carbon emissions, and regenerate oxygen by using vegetation filters to pass the air through green spaces with a purification intent. The use of different plants and green solutions increases biodiversity and the subjective aspects of the user experience, such as beauty, calmness, and serenity. In urban outdoor areas, air can be purified using green roofs, suspended gardens, vegetated walls, or existing (preferably) or newly planted trees.

Furthermore, it is important to minimise the use of energy and balance this by producing new energy locally, that is, by optimising the energy balance to a positive result with no carbon emission impact. A lower-energy consumption is required to meet the requirements of low-energy or zero-energy building standards or certification schemes. When employing such low- or ultra-low-energy requirements (less than 15 kWh/m^2 annually), it is essential to design the insulated envelope as airtight as possible, guarantee minimum fresh air renewal, and avoid overheating during the hot and warm seasons. The outer envelope walls tend to be quite thick due to the insulation requirement, that is, the sizing of facades and fenestration must be carefully considered to be suitable for the rest of the regenerative design. Passive solar gains, which promote energy savings, should be optimised on the south façades, together with flexible shading devices, to prevent overheating during the warm and hot seasons. Local rules of thumb should be used for the sizing of the window/wall ratio in all orientations. The use of passive cooling or heating by bypassing and exchanging stable temperature conditions in the soil surrounding the building is also favourable. When dimensioning a locally renewable energy system, estimations of the amount and size needed that exceed the estimations of building energy usage should be determined during the early design phase. When choosing the type of renewable system, e.g. thermal, geothermal, photovoltaic, or other systems, the sizing and spatial integration of the building system, form, and envelope must be properly considered. The area intended for the PV devices, orientation, and positioning must be studied and estimated. In addition to this study and estimation, the integration and sizing of the façade, roofs, or HVAC systems should be based on and adapted to the location of the building and the building site. The solar thermal system should also be adapted to the need for a hot water supply and stored in insulated tanks to meet user needs. The overall result of the balanced energy system should generate a positive regenerative impact outcome and ensure that the energy surplus produced locally will be integrated into the grid of public use.

For regenerative water usage and management, it is important to collect and harvest rainwater from different wastewater streams. From the perspective of regenerative sewage management, it is optimal to use in situ technology using plant-based purification or phytofilter systems which can treat both grey and blackwater. This solution also contributes to landscape greening and increases the biodiversity. A very important indicator of regenerative construction is a high and improved supply of water quality, that is, optimised water treatment and nutrient extraction from wastewater. The use of water tanks to collect rainwater during the wet period facilitates the independent use of water during the warm and dry seasons. Considering size and installation, it should be managed for each individual project, and special attention must be paid to supposed flooding occurrences, especially those predicted and caused by the future impacts of climate change.

The concept of design with nature introduces different configurations of vegetation, both indoors and outdoors. These solutions improve the quality of spaces and environmental quality, e.g. biophilia, biomimicry, regulation of humidity, and acoustics, and improve external quality, e.g. biodiversity and the heat island effect. Designing with nature begins by connecting green infrastructure with the building and its users to the ecosystem. Human well-being is based on a genetic connection to nature, and the use of biophilia approaches should verify this connection. It should be based on balanced nature-based solutions that connect flora and fauna to humans and promote variations in ecosystem services. This can include biodiversity, water management, urban food production, air purification, and human well-being and can also contribute to the recovery of the built environment from heat island effects, acoustic disadvantages, air pollution, and degradation of quality of life. Nature-based solutions include urban agriculture, green roofs, spaces, facades, trees, gardens, parks, ecological networks, and permacultures. Integrating these elements into the design of a construction project requires careful design and technical studies during the design phase, particularly in the early stages. Examples of special assumptions and considerations include root damage, artificial irrigation, structural overload, water flow and overflow, erosion, light penetration, solar orientation, consequences on envelope insulation and damp security, and plant diversity in order of presumed function and quality. Each construction project must address and integrate these issues individually owing to design objectives, location, site, and microclimate conditions.

The third strategy for designing a regenerative building involves the selection of building materials for the preferred design. This choice must be made without loss of biological or technical quality. Basic sources should comprise material declared in an environmental product declaration or in a third-party declaration of materials, such as by Cradle to Cradle (C2C). These sources of building material declarations should be used in conjunction with the aforementioned regenerative design principles. However, special attention must be paid to other common technical requirements, such as fire safety considerations, structural mechanics, and hygrothermal and acoustic performances. An important focus is to minimise embodied energy and carbon content. Materials from the technosphere, such as concrete, aluminium, or steel, may be used if they fulfil the requirements as far as potential for disassembly and reusable design and having a material certificate or declaration, as mentioned above. Of course, any toxic content of the material must already be excluded, as well as that within the production cycles of the material or parts of the material. The preferred materials are renewable and natural, i.e. wood, clay, straw, bamboo, or hemp. Special building construction elements, such as foundations, windows, specific safety devices, and technosphere materials, are unavoidable and must be declared.

The 12 design principles

When designing a regenerative building environment, defined criteria must be fulfilled and verified. To date, no clearly explicit criteria have been established for the regenerative design of a construction project; however, 12 regenerative design principles can be considered with various examples of assumed applications:

*   *Place and nature* – Ecosystem-place-based design approach for regenerating the ecosystem and making it possible for future development, i.e. ecosystem approaches, because each place or site is a unique dynamic entity; the building interacts with site environment and green neighbourhoods, ecosystem, and biodiversity protection; re-establishes lost or removed soil; and adapts to microclimate circumstances.

- *Energy* –Restorative and regenerative energy systems, effectively used and shared energy systems, and energy as part of a continuous restoration of ecosystems aiming to increase their quality, such as effectively used and shared renewable, restorative, and regenerative energy systems, and effective energy storage.
- *Carbon* – Carbon-neutral and climate-positive approaches are measures of reducing carbon, such as reducing a building's embodied carbon based on third-party lifecycle assessment sources.
- *Water* – Clean drinking water, such as improved water management and supply, and net positive water frequency supply by rainwater harvesting.
- *Material and resources* – Improving material and resource productivity through lifecycle assessment, material from renewable organic systems to conserve resources and maintain them for future generations, such as healthy and non-toxic materials, transparent labelling, the use of local renewables, responsible use, and conservation.
- *Waste* – Zero waste approach, design for disassembly, deconstruction, and flexibility of use, such as the C2C labelling system approach, upcycling, reuse, and recycling of materials and buildings.
- *Health and well-being* – Enhancing quality of life, reconnecting humans with nature, contributing to individual, community, and societal health and well-being without taking advantage of other people, the environment, or future generations, such as applying the principles of biophilia, ensuring access to healthy food, ensuring air quality, indoor air quality, daylight, comfort, and mindfulness.
- *Social equity* – Equality, gender equality, participation of people and countries, equity in the allocation of resources, inclusiveness of people and generations, support for vulnerable people, e.g. empowerment of women, older and young people, giving disadvantaged groups a voice, no threat associated with food production, and globally responsible action in dealing with resources.
- *Economy* – Regenerative, circular, crowdfunding, and sharing economy; sustainable production and consumption; collaborative business approach; place-based economies at multiple scales, such as restorative enterprises; redesign business models with a focus on selling products that create waste to provide services in closed-loop models; and energy unions.
- *Culture and community* – Address the social aspects of health, foster social cohesion and community identity, and promote accessibility and integration such as re-integrated and enjoyable heritage, buildings, and local heritage as visual, social, cultural, and economic catalysts for the community, accessibility of cultural and historical places, and inclusiveness of rural communities.
- *Education and inspiration* – Enable and encourage stakeholder participation, bottom-up cultures and initiatives, whole life education, increase awareness, such as encouraging pioneer movements, i.e. permaculture, urban gardening, and place-making, giving a voice to different sectors and interests of society, continuous and intergenerational learning and feedback, cooperation, interaction, and interdisciplinary planning.
- *Environment and mobility* – reduce carbon emissions caused by travel and transport, encourage walking and cycling, walkable and cyclable cities, rural–urban balance such as reducing transport volumes at the building site, encouraging and enabling car sharing, e-car charging stations, bicycle parking spaces and cycle paths, and footpaths to schools and shops.

Regeneration and circular economy

Regenerative design is based on anticipating the multifunctional evolution of buildings used in the future. In a rapidly changing society, buildings must be able to adapt to unknown changes and new socio-cultural and demographic issues. Therefore, it is essential to anticipate these changes and integrate strategies that allow the building to adapt to a variety of uses over time. At present, immense quantities of building materials end up in landfills or incinerators long before they have lost value, quality, or use. To begin, it is essential to define a logical choice in terms of constructive and structural systems, such as columns, beams, and slabs, to be able to upgrade reuse and cascade cycles thereafter. Flexible construction systems can make it easier to disassemble structures and recover, upgrade, modify, or remanufacture building materials. The selection of a flexible construction system allows future users to disassemble a building in terms of its elements, components, and materials to increase its resilience regarding multi-functionality and flexibility in spatiality and use. The modular design of construction systems allows for the reuse of structural components and materials, thereby increasing the multifunctional capacity of building uses, that is, modular construction systems that allow maximum spatial and user flexibility with parts that can be easily disassembled into reusable building materials and products. Examples of such implementations include timber, wood, metal, aluminium, concrete, and even masonry; modular structures (such as containers) or thin steel structures are other examples.

Designers must select a flexible construction system that allows for the combination of architectural elements and regenerative products. Such a system is a key to future modifications by the addition, subtraction, or replacement of the envelope and façade layers. For the chosen system, space analysis of the building must be performed to evaluate nature-based solutions and connections to ecosystem services. Depending on the microclimate condition of the building site and its geographical location, certain design elements may be more appropriate, thus improving indoor and outdoor air quality, water usage, and increasing the biodiversity, health, and well-being of end-users, enabling cultural and social cohesion and generating energy.

The final approach of the regenerative design framework is to address building products, optimise the material selection process, and integrate certified products into the building to increase its embedded material value. Each brick, board, and piece of wood or glass in a building has value. Instead of becoming waste, buildings must function as banks for valuable materials and decrease the use of natural resources to meet the resilience capacity of the planet. C2C-certified products or those with similar eco-labels are more beneficial for use because their connection to different cycles generates less biosphere or technosphere. Choosing regenerative building products and materials guarantees health, safety, and benefits to humans as well as to the environment. These materials or products are designed to be safely reintroduced into biological or technical cycles and are assembled or manufactured with 100% renewable and non-polluting energy. Regenerative building materials and products are designed to protect and increase clean water resources (as a basis for social and environmental justice). The use of such products also generates chain partnerships to validate the intermediary's recovery and reuse within the manufacturing process. This includes passing a passport for each material and creating a unique database for buildings to facilitate reuse in the future. Sustaining the value of materials is the key to the circularity of material use, and refining this value is central to regenerative buildings. Integrating material passports with a reversible building design to optimise circular value chains leads to a significant reduction in resource use. Tracing building materials and products will increase product lifespans and enable product and material reuse, recycling, and recovery. Furthermore, an upgraded cascading approach for recovered materials and products will reduce the generation of waste along the product chains in different manufacturing processes and decrease the utilisation of virgin materials, emissions, and depletion of harmful substances.

8.4 Regenerative construction

In a traditional construction project, the client would appoint architects and consultants to design and project managers to select the contractor, and depending on the contractual form, the contractor would procure subcontractors, material suppliers, and other services needed on site. However, this disjointed and adversarial approach does not enable regenerative construction. One approach for enabling regenerative construction performance is to establish early contracting procedures in the project. Contractor skills should be introduced in the early design process to bring regenerative sustainability and design in a performable construction project, and to introduce cost efficiencies during the preconstruction phase. The earlier the contractor is appointed in the design process, the greater the benefits they can bring to the project.

A great deal of project and sustainability value created during the design process of the project can vanish throughout the construction phase without close and focused project management. Accordingly, effective contract administration must be conducted throughout the regenerative construction project through project management. Documentation (e.g. progress reports, photos of progress of site works) must be carefully maintained. Effective and transparent communication should be maintained throughout the project management process so that communication errors and claims can be resolved without turning into disputes. In the initial part of the construction process, education (i.e. health and safety) should be provided to subcontractors and suppliers regarding sufficient sustainable and regenerative construction processes and requirements as well as onsite circularity management. Supplier risks, both environmental and social-related, and potential supplier improvements should be detected, and necessary precautions should be taken for the smooth flow of the work.

Procurement and contracting

The procurement phase of a regenerative construction project must be aligned with all sustainability criteria to be fulfilled, together with the regenerative scope of the construction project. The contract should enforce the contractor and subcontractors at all levels to perform as stipulated by the client, including their performance activities onsite, which reflects the level of requirements of the contractual work. Furthermore, this phase should also cover the risk, responsibility, and right allocation among the contracting parties with respect to the construction manager of the regenerative construction project. The risks should be allocated to the party that best controls them. All rights and obligations of the contracting parties should be covered and should not overlap. All statements in the contracts should be clear to avoid misunderstandings.

LLC or partnering can support regenerative construction projects as it supports sustainability. Partnering enables a cooperative approach among contracting parties, commitment to invest in green initiatives, and reduces the footprint of the supply chain. As partnering can be a solution for improving the central control of managing construction with high uncertainty and complexity, it can also support the dynamic, innovative, and iterative nature of regenerative construction. Partnering can be a tool for contractors to establish a supply chain with reliable subcontractors and suppliers who have a similar working culture and who comply with regenerative requirements. Furthermore, a subcontractor and supplier team under the partnering cooperative can support the learning process and knowledge sharing required for the performance of regenerative construction. Suppliers and subcontractors in regenerative supply chain management must have knowledge of agile construction project management to support the iterative nature of the design process. Iterations in the design process must be stopped at a certain stage because if they occur in the construction phase, they can cause different variations, resulting in increasing environmental footprints or other cascading impacts. Further, they can increase the final cost, for example, rework of performance.

Furthermore, regenerative construction contracts can cover all sustainability criteria, i.e. not only environmental but also social aspects of construction performance, such as ensuring ethical sourcing, prohibiting unethical sourcing activities and discrimination, and referring to Corporate Social Responsibility (CSR) or UN Global Impact commitments.

Standard contracts used, described as 'General conditions to the contract for construction' as well as company-specific contracts, must be updated. The contract documents should include in-house guidelines, environmental management systems such as ISO 14001, social standards such as CSR or UN Global Impact, and the company's compliance obligations or other codes of conduct.

As contracts at all levels need to cover all sustainability requirements, the contract documents should include the site waste or recycling management plan and the requirements of the suppliers and subcontractors to provide their staff with sufficient adapted education on sustainability and regeneration principles to improve their performance. The contracts should enable effective control mechanisms (e.g. various inspections to the production sites of suppliers as well as to the sites for subcontractors' performed works if not on actual construction sites) to ensure that the regeneration performance required is fulfilled. The contract should also require suppliers and subcontractors to respond to stakeholder enquiries, both external and internal, to detect and get feedback on their weak points regarding the environmental and social risks. Such action will also initialise a requirement for continual improvement of their part in the regenerative construction process. Effective contract administration throughout the construction and management of regenerative projects is essential.

Construction phase

The construction phase should begin with effective resource and site planning to support a smooth flow of material, information, equipment, and labour and to reduce non-value-adding activities such as waiting and transportation. Shared resources must be coordinated. Effective site planning for regenerative construction projects affects the sustainability performance of the construction work as well as the health and safety issues, for example, failing to weather-proof material could cause undesirable waste.

Effective site waste management and planning requires a well-structured management plan concerning how to reuse or recirculate, agreements between contractors and sub-contractors to determine who is responsible for managing waste onsite, procurement of waste companies offering beneficial waste segregation systems for recycling and disposal, and the appointment of a designated waste and recycling manager to coordinate the delivery and storage of materials.

To ensure that sufficient and necessary precautionary measures are taken, control activities should be conducted throughout the construction phase to identify and act on deviances and to take necessary precautions in a timely manner. Performed work needs to be controlled to ensure that it complies with the contractual design documents and the specified control mechanism (e.g. various inspections, quality verifications, and visits to the site). Control activities of precaution measures concerning embodied carbon and carbon emissions should be undertaken at the building product and supply-chain levels, complying with contractual requirements and with the contractors' own compliance obligations. Suppliers and subcontractors should be educated on how to reduce their carbon impact and other sustainability deviances in their performance.

Handover and commissioning

Handover documents of the regenerative construction project should be provided to the client by the main contractor when the stipulated contractual performance is completed. These documents include operator and maintenance manuals, spare parts catalogues, as-built drawings, and operational staff educational documents.

Operational manuals should include information on the supposed time between failures and to repair. To understand the supposed time of failure of construction systems, each system must be analysed and its failure, repair frequencies, and maintenance frequency determined. The supposed time to repair is the duration needed for repair in the case of a failure and the required maintenance duration. These estimations of durations need to be considered for the entire operational life of the regenerative construction project and its systems. These maintenance durations affect the operational cost and revenue during the building's life cycle, and an effective planned maintenance system is crucial to minimise unexpected repair measures due to unexpected failures. The planned maintenance system documents should be submitted to the client by the main contractor. Preparation of the planned maintenance system should include:

- Each specific piece of equipment and its expected durability
- System components and maintenance including:
  o the time schedule when each specific maintenance occurs
  o information on capacity
  o required spare parts
  o required duration with work hours and
  o qualification of maintenance personnel
- Documentation of vital equipment and systems considering operational requirements (especially for certain building types, e.g. hospitals)
- Equipment-specific standards
- Reparation of as-built or record drawing documents including operator manuals, maintenance manuals, spare parts catalogues, and training documents of the regenerative construction project

- Education and training for operators, maintenance staff, instructor documentation, and training warranty from the contractor. Sufficient education should be provided specifically to the operators, maintenance staff, and instructors before the commissioning phase at the end of the construction process. Sufficient verification of educational requirements should be included upon handover.

8.5 Regenerative operation and maintenance

A regenerative construction project must be considered a process that continually evolves over time. Regenerative development during the design process and the construction stage does not end with the delivery of the final drawings and approvals nor with the construction and handover of a project. As these projects are similar to living organisms, their regeneration performance should be consistently updated with contemporary technologies and, considering the dynamics of nature, comprehension of the state of the art of sustainability today will not necessarily be the same ten years from now. Regenerative development implies continually maintaining adaptive capacity and the capacity to renew knowledge. Therefore, throughout the building's life cycle, contemporary innovations and new equipment and systems should be analysed and thoroughly examined to monitor the impact of the ecosystem through ecosystem analyses. If the existing equipment is outdated compared to the new equipment, a schedule for exchanging or upgrading should be implemented. Sustainable facility management (SFM) and contractors should be responsible and defined in the contract between the client and the SFM contractor. Depending on the circumstances and contractual requirements, such as partnering or soft landings, the client or SFM contractor could pass this responsibility on to the construction contractor owing to their knowledge of construction performance, installation skills, or familiarity with the equipment.

If the installation of the equipment, for which space and infrastructure have been constructed beforehand, can be performed in the post-construction phase, the relevant supplier can perform the installation. Depending on the contract conditions, this can be coordinated by the client, the SFM contractor, or the original construction contractor. Both concepts can be supported by the design and construction allocated during the design process.

When maintaining and repairing based on the maintenance schedule defined in the contract between the client and the construction contractor, the contractor can coordinate maintenance work with the SFM or a relevant supplier, as constituted in the contract conditions. Moreover, a framework is needed to integrate social and cultural aspects into SFM practices, especially small- and medium-sized FM enterprises, to perform the regenerative development of operations and management (O&M).

To turn SFM into regenerative facility management, targets should be established for social, economic, and ecological regenerative development.

The latter ecological targets could include:

- Reduction of resources with a focus on circular economy
- Usage of recyclable building materials
- Consideration of disassembly and re-use of material
- Reduction of energy consumption and embodied carbon, and usage of renewable energy

- Reduction of space requirements
- Safeguarding the ability to maintain, re-use, and de-construct buildings
- Substitution of dangerous and hazardous materials impacting people or the environment, based on the precautionary principle

Economic targets for regenerative facility management could include:

- Building space optimisation for more efficient usage, using BIM and virtual reality digital technology to monitor effectiveness
- Optimisation of lifecycle costs involving different stakeholders to design, construct, handover, use, maintain, and re-develop buildings from a long-term, circular economy perspective
- Facilitating the most efficient O&M methods enhanced by digitalisation, ecosystem practices, and responsible procurement
- Using green bonds and crowdfunding as the basis of financial management

Social targets to facilitate regenerative development could include:

- Supply of a balanced number of buildings for work and life, and developing mixed-use and hybrid facilities in the context of urban regeneration
- Physical and psychosocial well-being, together with health, safety, and security requirements
- Identification of different social groups and social impacts – resilient buildings and neighbourhoods integrate different social groups and provide synergy
- Communication of regenerative values to users to increase the awareness of regenerative actions

It is important to identify the impact and opportunities of a regenerative approach to both construction and operations, in addition to the regenerative use of buildings. It is also essential to continuously optimise the benefits to the environment and users and to ensure that the initial quality is maintained or enhanced.

The steps towards regenerative Facility Management include:

1 From scheduled maintenance towards on-demand maintenance.
2 From recycling only towards self-sufficient solutions.
3 From a linear economy towards future life economy.
4 From human intelligence towards artificial intelligence.
5 From the passive user towards the active user.
6 From monitoring single indicators towards monitoring integrated indicators of co-operation and connections.
7 From maintenance by the service provider towards maintenance by the user and prosumer.

8.6 Regenerative rating and certification

Building certification schemes to achieve sustainability ultimately seek to advance the building sector towards ever-higher levels of improvement and transparency. Recent years have seen continued growth in the number of green/sustainable building certification standards, and more buildings are being certified. Globally, major certifications, such as LEED, BREEAM, Passivehouse, DGNB, and WELL, continue to be widely used.

Green buildings and sustainable building certifications focus on low and close to zero carbon emissions, and energy-efficient buildings are an important ingredient in achieving a low-carbon building stock. However, they must be based on common definitions alongside mandatory building codes and progressive policies. Many of these schemes have been implemented, certifying buildings that achieve net zero based on performance data, including standards from the national green building councils. Examples of net zero schemes include LEED Zero (USA), EDGE Zero Carbon (UK), NollCO$_2$ (zero carbon emission from Sweden GBC), DGNB International (Germany), and ILFI Zero Carbon (USA) certifications. The Green Building Council of Australia has released an updated Green Star tool, requiring that buildings be net-zero, fossil fuel-free, and 100% powered by renewables to score the highest possible six star rating.

A comparison of different rating and certification schemes reveals deviations in the priority setting of sustainability and environmental objectives. Major building certification schemes that have been on the market for more than ten years, such as BREEAM, LEED, and DGNB, mostly highlight environmental data. Specifically, they emphasise the energy performance of buildings. Such a narrow focus on the separate aspects of sustainability has also been critically examined. While criteria regarding water, energy, material, or indoor climate can be found in nearly all certification schemes, aspects such as managing climatic responsive design, advanced sustainability efforts, or neighbourhood impacts are significantly less frequent. Moreover, social criteria are rarely explicitly stated as indicators in the schemes, and their implementation is minimised due to the low grades that can be achieved. Primarily, they can be found in neighbourhood manuals or city rating schemes and not in ordinary building certification standards.

Regenerative options

Various comprehensive certification schemes that focus more on assessing buildings in use contain more extensive requirements in relation to environmental issues. The requirements imply not only that the construction project itself is sustainable but also that the site and context be restored and, in some cases, regenerated to their original condition and beyond. These certification schemes require integrated design objectives supported by verification, documentation, and simulation for energy performance and indoor air quality controls. One such standard is the Living Building Challenge (LBC) by the International Living Future Institute (USA), which differs from more common rating systems by evaluating actual performance after 12 months of use rather than the intended performance expressed in the design phase.

The LBC, as an example, does not contain any methodology, workflow, or tool to reach compliance but is structured around a set of imperatives that is supposed to be applied to either new or existing buildings, interiors, landscapes, and infrastructures. Considered to be one of the most rigorous and demanding assessment schemes, it requires buildings to be net-zero carbon and energy, not only guaranteeing energy supply via renewables onsite but also providing all its water supply and managing all its grey and black water onsite, in addition to guaranteed access to nature and the incorporation of biophilic design. An international scheme, its approach is considered sufficiently flexible to be applied in various climate zones and countries with their own characteristics and national preferences. However, in addition to the barriers to sustainability standards, such as cost, lack of knowledge and experience, and time to gather the necessary information required for accreditation to the scheme, regenerative building rating standards have been criticised for a series of other reasons. These include:

- Lack of clarity in relation to what is considered a positive or restorative contribution and how it can be assessed
- Being questionable in terms of their efficiency
- Having aims too ambitious to be adopted by designers without systemic thinking and ecological frameworks
- Raising issues related to the scale of the proposed solution which is normally too small to provide clear ecological benefits as well as feasibility in relation to how they can be inserted in the existing context of cities and neighbourhoods

One significant barrier related to the implementation of the LBC example is national codes and requirements, especially those related to water provision and discharge. In general, this is prioritised to a centralised supply where maintenance and control are assumed to be more cost-effective, making it difficult for construction projects to use onsite water and sewerage systems. Another barrier is water quality control, as many authorities, local as well as regional or national, require water entering building premises to be of drinking quality. Consequently, it could prevent greywater reuse for other purposes, such as toilet flushing or washing machines. With regard to energy supply, regulations often prioritise connections to the public grid, and there are no incentives for the installation of small and individual onsite renewable energy systems with surplus feedback to the grid. Barriers could also be connected to reasons related to regulatory bodies' risk aversion in relation to innovative approaches to design and construction in favour of existing and traditional practices. While stringent and outdated building policies can be a significant barrier for sustainable and innovative construction, obstacles can also include a lack of clear codes and directives that would help organisations attract investment in green technology businesses and, hence, develop the industry further.

The LBC and the 12 principles

The 12 principles of regenerative design are divided into categories such as place, energy, water, well-being, carbon, resources and waste management, equity, education, economics, culture and community, and the environment and mobility. For each category, some aspects of interest can be used to determine how rating tools and schemes can incorporate them. The categories mentioned and their regenerative goals assessed regarding the level of quantifiable coverage are as follows:

- Regenerative aspects related to place, culture, and community categories focus on the integration of the sites into local natural and urban communities. The specific objectives include the incorporation of the design principles of bioclimatic, biomimicry, biophilic design, regenerative land use options, and community connectivity.

    Concerning the LBC first imperative of the place petal, it focuses on environmental protection, that is, construction can only be undertaken in previously developed land, grey fields, or brown fields and must be sufficiently distanced from environmentally protected areas such as wetlands and old-grown forests, unless specifically connected to the protection of these sites. Furthermore, it protects prime farmlands but considers monoculture agriculture to be a suitable place for development. The local community agriculture aspect requires construction projects to incorporate areas dedicated to urban agriculture and/or medicinal plants and be kept, with two weeks of self-sufficiency for residential developments, unless the project's main purpose is to protect the land where it is being developed or restored to its original conditions. The biodiversity aspect focuses on the protection and expansion of existing natural areas and demands protection of biodiversity outside the project site through investments in approved conservation organisations. Regarding environment and mobility, construction projects must provide specific measures to encourage building environments that support walking, cycling, and reduce the use of vehicles onsite.

- Regenerative energy and water use imply net positive water and energy use. The definition of these categories can be derived directly from the LBC scheme, as this certification is leading the way to regenerative energy and water use. The LBC energy petal imperatives include energy + carbon reduction and net positive carbon which include requirements for reduction and dependence on renewable sources and development of decentralised and safe infrastructure. More specifically, new buildings must be 70% less energy-intensive than the equivalent baseline building type, with 20% being associated with embodied energy in construction. In addition, onsite renewable energy production must cover 105% of the annual energy use.

  The LBC water petal imperatives responsible for water use and net positive water require that water usage not exceed the carrying capacity of the natural and local hydrologic cycle. Additionally, the water footprint should be 50% lower than the regional baseline. This implies that the construction project must verify that it can supply the building with onsite sourced water for 12 months with monthly metering to verify the different supply sources. Moreover, the overall target water footprint must not exceed the natural capacity of the site at 50% of the regional baseline.

  This petal also requires onsite stormwater, greywater, and blackwater management, as well as onsite water harvesting from rain, underground, condensation, and surface sources, and recycling for grey and blackwater after onsite treatment for reuse. Connecting to the potable-water grid is only allowed if the municipality enforces it or if the local aquifer is inaccessible or contaminated. However, the connection to the grid can only be performed for potable usage within the carrying capacity of the site. Connecting to the municipal sewer grid is allowed, but the treated water must return to the site for usage as long as the municipality has biologically based treatment.

- Closely connected to the objectives of energy and water is a carbon category that targets zero carbon emissions, but not only in the use phase but also during the entire life cycle of a construction project and its outcome. Beyond the outcome of carbon emissions directly from the project life cycle, it is possible to use carbon storage techniques to increase the positive impact of the project on the climate.

- The well-being category of regenerative construction projects focuses on the comfort and health of the building occupants, similar to the WELL certification system, which is considered the most advanced rating tool for this category. The LBC petal Health + Happiness contains three imperatives: a healthy interior environment, healthy interior performance, and access to nature. The first transfers central operation control to occupants, guaranteeing access to fresh air and daylight for a minimum of 75% of all regularly used spaces. The second set of indoor air quality targets mainly refers to standards for ventilation and a set of indoor air quality tests in compliance with current local test method standards before and nine months after occupation, but no specifications are provided for indoor air temperature and humidity control. The third refers to promoting human–nature interactions through biophilic design approaches and making post-occupancy evaluations within six to twelve months of occupancy.

- The resource category defines regenerative resource management with a lifecycle approach, which includes responsible sourcing, transparent reporting of built-in materials, elimination of toxic materials, and inclusion of options for disassembly during design.

The LBC Materials petal includes five imperatives:

- Responsible materials require that a minimum of 50% of the building's timber and wood material from the construction project be FSC-labelled, salvaged, or harvested onsite during clearing the site for construction or to restore or maintain ecological function of the site. The remainder should derive from other low-risk sources with content, origin, and fair extraction processes proven and 80% of construction waste should be diverted from landfills. This is also a requirement for using local sources of materials
  - the red list prohibits the use of dangerous materials, that is, materials with the greatest impact on human and ecosystem health, such as asbestos, bisphenol A, chlorinated polymers including PVC, chlorofluorocarbons CFC and HCFC, PFCs, PCBs, and toxic-heavy metals such as lead, mercury, arsenic, cadmium, and volatile organic compounds.
  - responsible sourcing requires purchasing materials from approved suppliers or supply chains and that they be third party-labelled
  - living economy sourcing requires the use of materials derived a short distance from the project site
  - net positive waste focuses on durability, adaptability, and reuse of new materials to be able to deconstruct with adaptable reuse of materials and with re-integration to the nutrient loop at the end of material life. Diverting materials from landfills is required.
- The equity category targets building users through design and operation goals for inclusivity, accessibility, transparency, and investment in local/global communities as well as CSR programmes and cultural heritage.

  The LBC equity petal centres on promoting human-scale rather than vehicle-scale places in universal access and inclusion, focusing mainly on restricting the size and percentage of land use for car parking. It also focuses on creating places with universal accessibility (regardless of gender, age, socioeconomic class, etc.), including universal access to nature, as well as outdoor fresh air and solar rights, compared with biophilic design approaches. Finally, it focuses on equitable investment, requiring 0.1% of the total project cost to be donated to charity, and requests transparency and disclosure of business practices of the construction project team to promote an equitable society.

- The education category also targets building users and a wider community, with the objectives of wide participation and inclusion of education programmes.

  The LBC education category is included in the beauty petal, as the imperative of Education + Inspiration is aimed at providing educational materials about the operation and performance of the construction project to the occupants and the public to share successful solutions and catalyse a broader change. All projects must provide a project case study, annual open day for the public, and copy of the operations and maintenance manual. Moreover, all projects except residentials must provide a brief brochure describing the design and environmental features of the project, install interpretive signage to teach visitors and occupants about the project, and develop and share an educational website about the project.

  The other LBC imperative of beauty, a petal concerning qualitative measures which includes design that uplifts the human spirit, is Beauty + Biophilia, which implies connecting project teams and occupants with the benefits of biophilia and incorporating meaningful biophilic design elements into the project. Construction projects must be designed to include elements that address the inborn human/nature connection. Each project team must engage in a minimum of one whole day of exploration of the project's biophilic design potential. Exploration must result in a biophilic framework and plan for the project that outlines strategy and implementation of how the project will be transformed by deliberately incorporating nature through environmental features, light and space, and natural shapes and forms (a biomimicry approach). Moreover, the project will be transformed by deliberately incorporating nature's patterns through natural patterns and processes and evolved human–nature relationships. The project is to be uniquely connected to the place, climate, and culture through place-based relationships; a meaningful integration of public art; and containment of design features intended solely for human delight and the celebration of culture, spirit, and place appropriate to the construction project's function.

- Economics focuses on the integration of a project into the circular economy value chain. The link between the sharing economy and the built environment is also considered. No specific LBC petal refers to economics; instead, it is blended with the other petals, especially with the resource petal.

Regenerative indicators

Benchmarks for the defined regenerative indicators include quantifiable targets for certain regenerative aspects. In the case of energy, water, and carbon use, net positivity can serve as a regenerative benchmark. In other cases, no clear limit exists between the sustainability and regenerative objectives, such as to determine the regenerative comfort parameters. In addition, objectives for which the level of compliance cannot be quantified (e.g. in the case of the objectives of biophilic design and beauty) must use qualitative valuation and verification. Due to the various types of benchmarks, the relation of the strictness of requirements in the different assessment schemes to the regenerative indicators differs if quantifiable or qualitive evaluation is used, such as if the contribution of the requirement is unquantifiable (e.g. the LEED Integrative Process credit requires attempting an energy optimisation of the construction project, which cannot be translated to a quantifiable value). Another example is when the credit has lower benchmarks than the regenerative benchmark, e.g. in the LEED assessment scheme, the maximum points can be achieved by achieving a 50% improvement in the proposed building performance rating compared with the baseline. Finally, if the credit requires the same strictness as regenerative benchmarks (e.g. the LBC assessment requires net positive energy use) or it represents the most stringent values possible (e.g. according to WELL, PM2.5, concentrations less than 15 µg/m indoors).

However, LBC and WELL target stricter levels of comfort, material sourcing, and transparency than the more traditional sustainability rating tools, implying that they not only limit the negative effects of an artificial environment but also attempt to implement positive effects, such as improved health and productivity.

Regenerative economics is least covered in the different systems, but DGNB includes an alignment of circular economy principles. Other systems refer mainly to regenerative economic targets, including indicators requiring participation in the sharing economy (e.g. sharing community spaces with local communities or shared transport facilities). Regarding the depth of alignment with regenerative objectives, LBC targets positive impacts on all the regenerative objectives in four categories (place, energy, water, and carbon). In two other categories – resources and education – LBC encompasses all categories, but the requirement can be more comprehensive (e.g. participatory project development is only partially included). In the well-being and equity categories, some of the targets are not addressed in the system, and the economic category is not explicitly covered. Regarding the other rating tools, WELL incorporates all regenerative well-being objectives and partially covers only four other categories. DGNB and BREEAM incorporate all categories, but in most cases, not all aspects, and with less comprehensiveness. LEED does not encompass equity or education.

Andreucci et al.'s (2021) analysis of the five selected rating tools discussed above, namely, BREAM, LEED, DGNB, WELL, and LBC, regarding sustainability or regenerative content in the certification schemes, suggests that the schemes, based on their purpose (sustainability, wellness, and regenerative systems), are the defining factor of how they incorporate regenerative objectives. As expected, because it was developed specifically to address regenerative development, LBC is much more comprehensive in its incorporation of the identified aspects, addressing most issues highlighted in the academic literature. On the other end of the scale, WELL is more narrowly focused, as it provides healthy buildings for occupants, but by incorporating the widest range of health-related regenerative aspects possible. In the middle range, the main sustainability-oriented rating tools, such as LEED, BREEAM, and DGNB, provide good coverage of the regenerative objectives that originated from traditional sustainability objectives, but with limited coverage of education and equity. Among the three latter systems, LEED does not address equity and education, nor does it set comprehensive targets in the other categories.

8.7 End remarks

The construction sector is a main contributor to carbon emissions and rising climate change but is also largely dependent on product design, the outcome of environmental impacts caused by climate change, and societal development due to this change. The recent IPCC (2022) report entails major impacts caused by the construction sector and what is needed to mitigate them. Roadmaps for the 2030–2050 period suggesting ways to mitigate or enhance the decrease of climate change impacts are more or less profound. However, when turning negative impacts into positive regeneration via the construction sector, what to do beyond point zero is a new approach. A condensed roadmap for a regenerative construction project is provided through the issues compiled from this book. Lastly, a section concerning the key conclusions from this textbook is presented.

Climate change impacts caused by the construction sector

As the IPCC (2022) in Mitigation of Climate Change states, the need to reduce all carbon emissions is extremely urgent as it is 'likely' that global warming will exceed 1.5 degrees during the 21st century and the 'likelihood' of limiting temperature acceleration to 2 degrees depends on rapid mitigation after 2030. Urban areas are likely to create opportunities to reduce energy consumption and carbon emissions through a systematic low-energy transition towards net zero emissions and beyond. However, this is only possible if emissions are reduced within and outside the boundaries through supply chains, which will also trigger beneficial cascade effects for other sectors. The reduction may encompass the storage and uptake of carbon, changing energy, and material consumption, and implementing new ways to produce electricity in the urban environment for individual buildings. When using nature- or ecosystem-based solutions, carbon uptake and storage can be achieved by using bio-based construction materials, permeable surfaces, green roofs and facades, trees, green spaces, rivers, and lakes.

In 2019, buildings accounted for 21% of total carbon emissions globally. Fifty-seven percent of these emissions were caused by generating offsite electricity and heat to buildings, 24% were produced directly onsite, and 18% came from embedded emissions. Emissions produced onsite are defined as fossil fuel, gas, or biomass-based combustion from cooking, heating, cooling, and hot water. Embodied emissions derive from extracting, producing, transforming, transporting, and installing construction materials or products in buildings. Several co-benefits can be achieved when energy consumption and carbon emission impacts from buildings are reduced, including health benefits due to improved indoor and outdoor conditions, productivity gains in non-residential buildings, new job opportunities in the local market, and improvements in social well-being. These benefits may be more valuable in the long run compared to the economic gain from specific energy savings. The vulnerability of buildings to the changing climate – such as temperature rise and increase in energy consumption leading to higher carbon emissions, sea level rise, increased storms, and rainfall impacting building structures, materials, and products – must also be considered. Such occurrences could result in increased energy consumption, increased client expenditures owing to higher maintenance or renovation needs, or the need to install or renew building equipment. Preventive action includes implementing well-planned energy efficiency and onsite renewable energy production to increase building resilience to these climate change impacts and reduce the need for climate adaptation.

Well-designed and effectively implemented actions for new buildings and renovated existing buildings have significant potential to achieve the SDG goals and reach zero emissions by 2050 as well as to facilitate adaptation to future climate conditions. Design approaches should integrate concerns, including building typology, form, and multi-functionality, to allow repurposing of unused building areas, meet clients' and occupants' changing needs of functionality, and avoid using carbon-intensive material and additional land. From a lifecycle perspective, low-emission materials, highly efficient building envelopes, renewable energy solutions, and nature-based solutions can be implemented. During the operational phase, using efficient building equipment, optimised use, and reusable materials and products can be considered. Renewable energy sources include solar PVs, small wind turbines, solar thermal collectors, and biomass boilers. It has been estimated that implementing such actions by 2050 could reduce more than 60% of carbon emissions from buildings globally; however, the decade of 2020–2030 is critical to accelerating knowledge transfer and know-how.

A condensed roadmap for a regenerative construction project

A condensed roadmap to initialise and perform a regenerative construction project is as follows: The initial part can be used to create a restorative version, and then, through continual improvements, it can be finalised to propel projects towards a regenerative end result.

- To begin, the story of the place, the site, and its surroundings should be compiled and include existing and past stories. It can be assessed with the help of ESA and the LBC place petal.
- Client commitment of a regenerative learning process with continual improvements is crucial and extends the construction project beyond a finished product, the building, towards a lifecycle commitment. A prerequisite is an existing EMS system at the client's organisation to work this commitment through and to apply the commitment of continual improvements throughout the life cycle of the project, that is, the building's life cycle. The client must also allow a wider timeframe in the early design phase to engage all stakeholders in commitment.
- Create a core team to continuously facilitate long-term iterative cycles of clients, project teams, and stakeholders, and implement the team in the project organisation as an impartial control function and extend this function throughout the life cycle of the building as long as possible.
- Capacity building or knowledge transfer is very important for all stakeholders, especially clients and the project team. This can be achieved using the client's EMS system. Other important stakeholders and financing bodies should also be included in capacity-building schemes.
- The project design must shift to non-linear, cyclic, process-based descriptive continuous monitoring, iterative, and integrative design processes. In addition, it should use bioclimatic (which is already mainstream), biophilic, and biomimicry for nature-based designs, and a design that includes reused and reusable materials and products, preferably with optimised cascading effects, and the use of the LBC beauty petal. The design should meet the results and restorative and regenerative suggestions based on the story of the place.
- It is important to choose a contractor as early as possible during the design phase of a construction project. LCC or partnering contracts could be suitable solutions with stringent, but adaptable requirements. In contractual documents, requirements should be as stringent as those of subcontractors, material suppliers, and the main contractor throughout the supply chain. Effective contract administration throughout the construction and management of regenerative projects is essential. It is also essential to establish a supply chain with reliable subcontractors and suppliers that have a similar working culture and comply with the regenerative

requirements of the client. The learning process and knowledge sharing required for regenerative construction must be supported. Suppliers and subcontractors in regenerative supply-chain management must have knowledge of agile construction project management to support the iterative nature of the design process. Construction performance must have controls and ensure the handover with sufficient documents.

- O&M must contract with SFM which should be converted to regenerative facility management.
- When considering everything, the scheme of LBC certification is a good path towards progress and is principally the only rating scheme for regenerative buildings although it does not encompass all 12 principles of regenerative design.

Further on to a construction process for a generative future

The actions needed to decrease the risk of severe climate change at the end of this century will depend on what is done in the next 30 years. It is vital to reach a level of carbon emission reduction that enables it to withstand global warming above 1.5°. For the construction sector's share of carbon emissions, which is substantial, knowledge transfer to all stakeholders must be in focus. The IPCC AR6 WGIII (2022) report states that 2020–2030 is critical for accelerating this knowledge transfer. Several roadmaps exist to move the built environment and construction towards enhancing the development of sustainability and reaching a point zero for carbon emissions. Gradually, step-by-step – through the roadmaps leading to 2050, achieving zero, and then beyond zero – positive impacts will slowly increase through continual improvements.

Furthermore, to be proactive, the construction process must change from a linear process to a regenerative one, including the circular economy approach. A regenerative construction process must be extended to the product, building, and entire life cycle, including regenerative facility management and circular material productivity. The main drivers responsible for such development are clients, investors, and policymakers. Knowledge must be transferred to designers, contractors, and operation and maintenance organisations, and the contractors must be assigned in the early stage of design with proper contractual agreements. Continuous communication with stakeholders must be maintained in the early stage of the design phase. The design teams should be very familiar with eco-engineering knowledge of ecosystem services, as well as with social anthropologists and clients, forming a core team to assess the entire regenerative process.

Parts of such processes already exist on the market as mainstream construction requirements, but without the whole perspective. Although we see a forest, with its individual trees, we do not yet see the overall forest in its entirety from afar. We absolutely do not see other forests and are barely aware that something more is out there.

Only a few construction projects will reach the level of a regenerative construction process; most future projects will reach a restorative level. However, as long as the outcome is beyond zero and the impact is positive, it is beneficial enough to regenerate a great deal of biodiversity and the natural ecosystems that were lost during human resource consumption that did not consider our environment's limited level of resilience and the Earth's well-being. A paradigm shift in the construction sector is necessary in the coming decades to recognise that the whole is much more than just its parts.

Bibliography

Andreucci, A.M. et al. (eds.) (2021) *Rethinking Sustainability towards a Regenerative Economy, Future City*, Vol 15, Springer Nature, Cham, Switzerland, 2021.

Attia, S. (2018) *Regenerative and Positive Impact Architecture – Learning from Case Studies*, SpringerBriefs in Energy, Springer Nature, Cham, Switzerland, 2018.

Baper, S.Y. et al. (2020) *Towards Regenerative Architecture: Material Effectiveness*, International Journal of Technology 2020, 11(4), 722–731.

CIOB (Chartered Institute of Building) (2014) *Code of Practice for Project Management for Construction and Development*, 5th edition, Chartered Institute of Building, John Wiley & Sons, London, UK, 2014.

Cole, R.J. (2020) *Navigating Climate Change: Rethinking the Role of Buildings*, Sustainability 2020, 12, 9527.

Collins, D.A. (2019) *Green Leasing – A study of barriers and drivers for Green Leased office in Norway*, Doctoral Thesis at NTNU, 2019:150, Norwegian University of Science and Technology, Trondheim, Norway, 2019.

Forsberg, M. and Bleil de Sousa, C. (2021) *Implementing Regenerative Standards in Politically Green Nordic Social Welfare States: Can Sweden Adopt the Living Building Challenge?* Sustainability 2021, 13, 738.

Gibbons, L.V. (2020) *Regenerative—The New Sustainable?* Sustainability 2020, 12, 5483.

Haselsteiner, E. et al. (2021) *Drivers and Barriers Leading to a Successful Paradigm Shift toward Regenerative Neighborhoods*, Sustainability 2021, 13, 5179.

IPCC (2022) Summary for Policymakers. In Climate Change 2022: Mitigation of Climate Change Contribution of Working Group III to the Sixth Assessment Report of the Intergovernmental Panel on Climate Change [P.R. Shukla, J. Skea, R. Slade, A. Al Khourdajie, R. van Diemen, D. McCollum, M. Pathak, S. Some, P. Vyas, R. Fradera, M. Belkacemi, A. Hasija, G. Lisboa, S. Luz, J. Malley, (eds.)]. Cambridge University Press, Cambridge, UK and New York, NY, https://www.ipcc.ch/report/ar6/wg3/, access 2022-04-10.

Peretti, G. and Druhmann, C.D. (eds.) (2019) *Regenerative Construction and Operation, REthinking Sustainability TOwards a Regenerative Economy, RESTORE*, Working group three publication, Eurac Research, Bozen, Germany, 2019.

Reed, B. (2007) *Forum: Shifting from 'Sustainability' to Regeneration*, Building Research & Information 2007, 35(6), 674–680.

Sertyesilisik, B. (2017) *A Preliminary Study on the Regenerative Construction Project Management Concept for Enhancing Sustainability Performance of the Construction Industry*, International Journal of Construction Management 2017, 17(4), 293–309.

UNEP (2020) 2020 *Global Status Report for Buildings and Construction: Towards a Zero-emission, Efficient and Resilient Buildings and Construction Sector*, United Nations Environment Programme Nairobi, Kenya, 2020.

Index

Note: *Italic* page numbers refer to figures.